THE ICE-AGE HISTORY OF SOUTHWESTERN NATIONAL PARKS

THE ICE-AGE HISTORY OF SOUTHWESTERN NATIONAL PARKS

Scott A. Elias

Smithsonian Institution Press

Washington and London

Copy Editor and Typesetter: Princeton Editorial Associates
Production Editor: Jenelle Walthour

Library of Congress Cataloging-in-Publication Data
Elias, Scott A.
The ice-age history of southwestern national parks / Scott A. Elias
p. cm.
Includes bibliographical references and index.
ISBN 1-56098-679-4 (pbk. : alk. paper)
1. Geology, Stratigraphic—Quaternary. 2. Geology—Southwest, New.
3. National parks and reserves—Southwest, New. I. Title.
QE696.E45 1997
560′.45′0979—dc20 96-15415

British Library Cataloguing-in-Publication Data is available

Manufactured in the United States of America
04 03 02 01 00 99 98 97 5 4 3 2 1

The paper used in this publication meets the minimum requirements of the American National Standard for Information Sciences—Permanence of Paper for Printed Library Materials ANSI Z39.48-1984.

For Nancy, who shares my love of the deserts and canyons

CONTENTS

FOREWORD

This book is about new discoveries of our *wild West.* The West of which I speak is much older than Wyatt Earp or Kit Carson. It long predates Coronado and Cabeza de Vaca, who came and went in the 1500s. Our author and guide, Scott Elias, takes us into the last ice age over 10,000 years ago, explaining how to "time travel" into a time warp little known until recently.

Scott introduces us to the basic field techniques, including dating methods for the kinds of deposits that yield fossils. He discusses the new field technique of prospecting for packrat middens, the encyclopedia-sized time capsules found in dry caves. The packrats collected plants growing outside their caves or rock-shelters. In addition, these animals placed the hard parts of insects, bones of vertebrates, and even the scales of lizards in their middens. From the middens we learn of unexpected changes in the range of plants and animals in the Southwest during the last ice age, and afterward.

The Southwest Scott shows us goes beyond fiction, for truth, as we all know, is stranger than fiction. Using some of the West's finest national parks as his stage, and taking us outside them when it suits his purpose, Scott introduces us to the extinct mammoths and mammoth hunters. He reports climates once cool enough to replace ponderosa pine trees with limber pine or spruce and desert cacti with juniper and oak. We learn of caves so dry that the bones of extinct ground sloths are still covered with tissue and hair. The ancient nests of extinct condors still contain eggshells and feathers!

The first Americans took part in or witnessed all this and much more. They succeeded in farming a dry land, risking everything and sometimes losing to deadly droughts. Although climatic change can be blamed for prehistoric abandonment, Scott does not overlook evidence of fuel wood depletion as well. When people gamble with their carrying capacity, they risk disaster. The cause is always both natural and man-made.

I won't delve much deeper into the mysteries, for that is Scott's job. But as one might guess, the new findings are beginning to change the way in which we may think about the West before it lost its wildness. Scott Elias writes of the loss of an American Serengeti. What does he mean by that?

American Serengeti is his name for an untamed land of large herds of large animals, the native American megafauna. Mammoths, extinct camels, extinct horses, and extinct bison were the main species. There were dozens more. Such animals had long existed in the New World before people arrived over the Bering Bridge 10,000 years ago. I hesitate to put a firmer date on the arrival because, although all archaeologists would agree that people were certainly here by Clovis time, 11,000 radiocarbon years ago, they seek even earlier arrivals.

And why not? There is a strong ecological case to make against expecting any people to have lived in America before the Clovis foragers. A New World sustaining the numerous extinct animals Scott mentions should have supported many people, living happily on the wealth of plants and animals nurtured in turn by a favorable climate and, as in the Midwest, by deep, rich soils. In particular, if some slow-moving species such as the ground sloths, giant armadillos, and large land tortoises were ridiculously easy to hunt, as we have every reason to expect, they should have disappeared much earlier than they did. Instead they vanish coincidentally with the time of the Clovis hunters, along with the mammoth.

What happened? Apparently we are wedged between several questions that continue to puzzle the experts, or at least any consensus of experts. When (before Columbus) was America discovered? When did the mammoths and other large animals become extinct and why? Some would answer these questions as a unit. Others fear to tackle more than one at a time.

Whatever the answers to be found here in Scott's book, it is clear that our view of the natural landscape of the West must be purged of a bias or two. Most Americans have no idea that mammoths, horses, camels, and others not mentioned in "Home on the Range" were actually native here. Most Americans have not the slightest idea that the cooler climates of the last cold stage, or "late Pinedale" as Scott calls it, are typical of the last million years, not the exception.

If the West of Lewis and Clark and the fur traders, replete with buffalo, elk, and pronghorn, was *wild*, the West of the mammoth hunters over 10,000 years earlier can only have been *wilder*. If so, our history as now taught—covering the last 500 years only—is seriously flawed. Such shallow history sells the West short and no patriotic American should stand for it!

We need to slough off the slush of the last few centuries and the alluvium of popular opinion, as Thoreau recommended in *Walden* 150 years ago, to dig deeper and anchor in bedrock—a *point d'appui* he called it. Before the mammoth hunters, before any Americans, "natives" or "latecomers," we have the game-rich American Serengeti, glacial tongues in the Rockies, deep lakes in the Great Basin, limber pine—not piñon—in Canyonlands, and oak and juniper—not creosote bush and ocotillo—in the Big Bend. Without hesitation, we can now use the superlative: Scott Elias's book introduces us to the *wildest West!*

This is deep history on nature's timetable. Come, see for yourself! And keep an eye peeled for the shadow of ground sloths at sunset! The land has not forgotten them.

Paul S. Martin
Emeritus Professor of Geosciences
Desert Laboratory
University of Arizona

ACKNOWLEDGMENTS

I thank the National Park Service Publications Office in Denver for financial support that allowed me to start writing this book. I thank the U.S. Geological Survey Photo Library in Denver for access to and permission to use several photographs of geologic features in the parks.

The faculty of the Institute of Arctic and Alpine Research, University of Colorado—including Nelson Caine, John Hollin, Susan Short, Tom Stafford, and Mort and Joanne Turner—made helpful suggestions on first drafts of chapters and figures and also provided photographs. Paleontologist Elaine Anderson of the Denver Museum of Natural History provided Pleistocene faunal lists for the park regions.

Paleoecologists Paul Martin and Julio Betancourt of the U.S. Geological Survey, Tucson, and anthropologist Linda Cordell of the University of Colorado Museum reviewed all or part of the first draft and made many useful suggestions.

Finally, I thank my family for their support throughout this project and for doing without me through many evenings and weekends.

In addition to the support of the National Park Service, other financial support for the preparation of this book was provided by a grant from the National Science Foundation to the University of Colorado for Long-Term Ecological Research (LTER), DEB-9211776. Support for my paleoecological research in the Southwest has been provided by grants from the National Science Foundation and the National Park Service.

PART ONE

Paleoecology

Why We Need to Study Past Ecosystems

THIS BOOK DEALS WITH the prehistoric environments of the American Southwest. The term *prehistory* covers several billion years on this planet. It is a time frame we all find difficult to grasp, even if we are well acquainted with some of the more spectacular episodes in prehistory, such as the age of dinosaurs. But having gotten that far, can we honestly say that we have a good understanding of what a million years means, or what percentage of total Earth history it represents?

It is somewhat easier to think about the events of the last few tens of thousands of years. It was during this time period that the glaciers of the last ice age advanced and then began to melt. It was also during this time that humans moved into North America from Asia, and, more recently, from Europe and Africa. We have come to learn something about that time, when large parts of this continent were clothed in primeval (old-growth) forests or unbroken expanses of tall-grass prairie.

In an attempt to preserve some of that untamed wilderness, our government has set aside tracts of land as national parks, beginning with Yellowstone in 1872. Often these parks represent the last, best examples of entire **ecosystems** close to their primeval state. However, even Yellowstone is ecologically unbalanced. For example, there are far too many elk in the park, but there is a lack of predators (human and otherwise) to keep their numbers down. Perhaps the introduction of wolves will help redress the balance, but human hunters (Indians) were a part of the ecosystem for 10,000 or more years before the creation of the park. Who will take their place in the ecosystem?

National parks in the American Southwest also hold on tenaciously to pieces of ecosystems, keeping them out of the hands of commercial developers, ranchers, and others whose plans might radically alter the character of the ecosystem. Big Bend National Park contains one of the largest tracts of Chihuahuan Desert not in private hands. Although the park region was freed from cattle grazing as recently as 1944, the natural vegetation has made a surprising comeback in the last few decades. Canyonlands and Grand Canyon national parks also preserve large regions of relatively undisturbed habitats on the Colorado Plateau, ranging from Mojave and Great Basin deserts to montane forests.

Recent archaeological research suggests that the North American ecosystems first viewed by Europeans in the fifteenth century were, in many ways, significantly altered by Indians. For instance, Indians probably set fire to large tracts of prairie on the Great Plains, altering the abundance and diversity of the vegetation. A growing body of fossil and archaeological evidence indicates that southwestern landscapes were also considerably altered by Indian activities. We need to understand the role of humans on the landscape, past and present. One of the best ways of approaching this problem is through the study of the fossil record, and the reconstruction of how ancient plants, animals, and people interacted.

My aim in writing this book is to provide an overview of this more recent period of prehistory for the American Southwest and of the methods used to reconstruct past environments. The book draws from the work of many scientists. I have brought together information from studies of fossil plants and animals as well as geological data, all of which can be used to help reconstruct ancient climates. Finally, I provide an overview of the early peoples of the Southwest, especially the first nomadic hunters that migrated south from Alaska and Canada. This book is an attempt to fill the gap between available books about modern ecosystems and those concerning bedrock geology. Obviously, there is a large interval between the events of millions of years ago and the events of the last hundred years. The first step in bridging that gap is providing information on the hows and whys of paleoecology. Then we will examine the ancient environments, vegetation, animals, and peoples of the Southwest.

Paleontological discoveries made in recent years in the American Southwest are forcing biologists, natural historians, and archaeologists to abandon long-held theories. The primary message derived from the fossils is this: the present is not the past. The deserts, woodlands, and montane forests of today were not always the way they are. In fact, during the last glaciation, the whole region worked differently, as dictated by radically different climates. In the Southwest, as elsewhere, the apparent long-term stability of modern ecosystems is really an artifact of the brevity of scientific observation. The fossil record demonstrates quite clearly, in a blow-by-blow account, that modern biological communities are just the latest reshuffling of plants and animals in an ever-changing progression, choreographed by changing environments. Modern biology fails to see these changes, because they take place on a time scale of centuries to millennia. Perhaps if human beings lived as long as bristlecone pine trees, ecologists would have a very different view of ecosystems. The cherished life zones described in detail by ecologists during the last hundred years would be seen as stairs on an escalator, constantly moving up and down the slopes of the landscape rather than firmly fixed to one elevational band.

Some biologists continue to prognosticate about how long species X has been living in piñon-juniper woodlands of New Mexico, based on its molecular genetics or its physiological tolerances as worked out in a laboratory. To be sure, these are interesting tools for the development of our understanding, but they do not provide the best evidence for figuring out how long species X has been living where it does today. The best evidence for that longevity probably comes from the fossil record. Where else can one find tangible proof of the comings and goings of a species through long periods of time?

I have attempted to express the ideas in my narrative as clearly as possible, using ordinary English words rather than scientific jargon. Many words used in science may be unfamiliar to nonscientists. I have tried to keep these to a minimum, but in some cases their use is unavoidable. When I use such words in this book, they are printed in boldface. I try to explain them as they are first introduced in the text, but all boldfaced words are also defined in the glossary at the back of the book. For instance, I could not avoid using the term **paleoecology** to refer to the science of reconstructing interactions between prehistoric plants and animals and their physical environments.

Chronologically, this book covers the *late* **Quaternary Period:** the last 125,000 years (Fig. I.1). During this interval, ice sheets advanced southward, covering Canada and much of the northern tier of states in the United States. Glaciers also crept down from mountaintops to fill high valleys in the Rockies and Sierras. Although glacial ice did not cover the American Southwest, regional climates were greatly affected by global cooling during glaciations. In addition, the ice sheets that covered more northerly regions forced changes in atmospheric circulation. It is thought that the **jet stream** split in two as it met the obstacle of ice sheets that were perhaps several miles thick. One branch flowed north of the ice sheets; the other branch flowed south, bringing increased moisture to the Southwest.

The late Quaternary interval is important because it bridges the gap between the ice-age world of prehistoric animals (some of which are now extinct) and

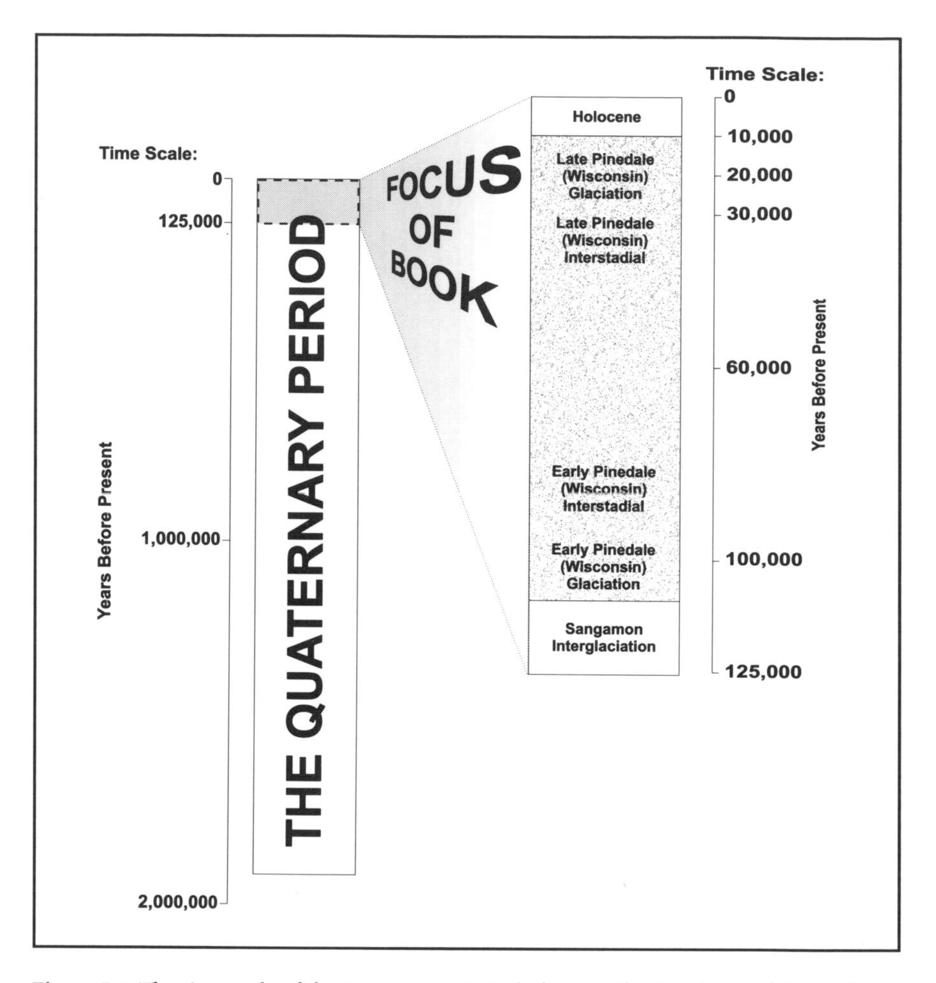

Figure I.1. The time scale of the Quaternary Period, showing the time interval that is the focus of this book: the late Quaternary, spanning the last 125,000 years.

modern environments and biota. It was a time of great change, in both physical environments and biological communities. It was also the time when human beings "came of age" in the world; they spread across the continents and started to become a major factor in the world's ecosystems.

The Pleistocene fossil record in regions ranging from arctic tundra to temperate grasslands is mostly preserved in waterlogged deposits of peat and lake sediments. This statement might lead you to believe that the fossil record of the American Southwest is extremely poor and that fossil sites are few and far between. Paradoxically, nothing could be further from the truth! The deserts have their own ways of preserving Pleistocene fossils. It turns out that the lack of waterlogged sediments is not a problem. In fact, fossil preservation in a hot, dry climate is sometimes far better than preservation in soggy sediments. Although an abundance of water may preserve some kinds of plants and animals in sediments, damp sediments that are exposed to oxygen scarcely preserve any fossils at all, because they provide the ideal conditions for decomposition. For instance, an animal carcass left on damp soil will rot in a few weeks; the same carcass left in a drying oven will lose its water, thus retarding decomposition. A pile of animal bones, hair, and dung may last tens of thousands of years in a dry cave. Another ingredient contributing to the success of fossil preservation in the American Southwest is the abundance of shallow caves and rockshelters. For example, the sandstone outcrops that weather into beautiful arches, pillars, and canyons also form numerous small caves. These caves have been home to many generations of animals, from the rock squirrel of today to the ground sloth of the last ice age. When animals die in these caves, their bodies are preserved, sometimes nearly intact, for many hundreds or thousands of years.

The final ingredient in the southwestern fossil preservation kit is the packrat. This industrious rodent collects food plants and other items and stores them in its den. Over great spans of time, dried plant remains, insects, and bones become cemented into black, tarry masses known as **packrat middens.** In dry caves, these middens can last tens of thousands of years, preserving smelly time capsules of ancient desert life.

This is the exciting surprise about the prehistory of the American Southwest: the region is loaded with fossil remains from plants and animals that have lived during the last 40,000 years. The parks and monuments discussed in this book are the repositories for whole menageries of extinct animals; they also have preserved a marvelous assortment of fossil plants. It is no exaggeration to say that the fossil records preserved in Southwestern parks tell us more about the prehistory of this region than we have been able to piece together for most other regions of the United States. And the fossils have a fascinating story to tell.

The Southwestern deserts we know today essentially did not exist in the past. During the last ice age, coniferous woodlands covered most of the Southwest; desert plants were scattered far and wide, living in pockets of arid climate. Many species of large ice-age mammals roamed the American Southwest, ranging from the majestic (such as the Columbian mammoth) to the bizarre (such as the giant ground sloths). At the end of the ice age, human beings entered the Southwest. Archaeological evidence for the Clovis and Folsom cultures was first discovered here. The end of the last ice age, about 11,000 years ago, brought down the curtain on this fascinating tableau. Many of the creatures have departed the stage forever, becoming extinct. Others, especially the plants, have shifted territories dramatically since then. Only the rocky outcrops that dot the Southwestern landscape have remained intact; everything else has changed. This book presents the story of the mighty, all-encompassing change that took place at the end of the last ice age, as well as an account of what went on in the Southwest before and after that transition.

Knowledge of late Quaternary ecology provides important clues toward understanding modern ecosystems, because modern ecosystems are the direct result of these past events. To try to understand present-day environments without a knowledge of their history since the last ice age would be like trying to understand the plot of a long novel by reading only the last page.

You will meet few unfamiliar plants and animals in this book. With notable exceptions, the flora and fauna discussed here are still growing or cavorting around on the North American landscape. The ones that have become extinct since the ice age are a fascinating story unto themselves, one that we shall explore. Fossil studies show that the modern ecosystems did not spring full-blown onto the hills and valleys of our continent within the last few centuries. Rather, they are the product of that massive reshuffling of species that was brought about by the last ice age and indeed continues to this day.

The study of past ecosystems is really just a form of detective work. A police detective reconstructing a crime has to reconstruct the following aspects of the case:

1. What happened?
2. Who did it?
3. How was it done?
4. When was it done?
5. Where was it done?
6. Why did it happen?

Apart from the last category, paleoecologists are basically saddled with the same questions. You might say that in ice-age fossil studies, the trail of clues has grown exceedingly cold. Certainly, the suspects and witnesses are all long dead. Nevertheless, the task, although difficult, is not impossible. Moreover, it's often exciting and challenging. And, as we shall see, it is becoming more important all the time.

In order to reconstruct *what happened* in the Quaternary, we need to look at both physical and biological data. The data from the physical environment include such information as types of sediments deposited and rates of deposition, types of landforms associated with glacial and near-glacial (called periglacial) environments, and changes in lake levels. The fossils themselves may suggest *who did it* and often show us *where it happened,* although we are often looking for external factors that influence the environment, such as changes in the amount of incoming solar radiation (called *insolation*). Other aspects of the physical environment make occasional appearances on the "who done it" list. These include volcanoes, earthquakes, and even the odd meteor or asteroid. The dating methods outlined in Chapter 3 provide information on *when it happened.*

Perhaps the trickiest question of all is "*How was it done?*" In other words, how have the various elements of the physical and biological world interacted on the global stage during the Quaternary? This question is the most difficult to answer, but it is also perhaps the most important, because we desperately need to know how the biological world responds to changes in the physical environment. We are inflicting our own changes on the environment at an ever-increasing rate. Overhunting, overfishing, pollution, and destruction of natural habitats have already wrought havoc on most ecosystems and have caused the extinction of untold numbers of species within the last few centuries. At present we find ourselves in the position of trying to understand how to sustain the remaining flora and fauna just as human activities place the natural world in ever greater jeopardy. Virtually our only means of gaining a greater understanding of how regional biotas respond to environmental change is to examine how they have responded to past changes. The current environmental crisis thus takes paleoecology out of the arcane realm of satisfying the intellectual curiosity of a few college professors and places it squarely in the middle of worldwide efforts to save our remaining ecosystems and their biota.

In this volume, I focus on the prehistory of five national parks and national monuments in the American Southwest (Fig. I.2). I chose these sites for two reasons. One is that, as nature preserves, the parks offer scientists excellent opportunities for research in many fields, including paleontology and geology; hence, much is known about these sites. Since the parks are relatively pristine, it is easier to compare past and present ecosystems there than in regions that have been greatly modified by human activities. I have worked on paleontological projects in Big Bend, Canyonlands, and Grand Canyon national parks and have gained familiarity with their ancient history. Another reason for my focus on the parks is that many people are interested in them and in the ecosystems they represent.

In the last paragraph, I used the term *pristine* to describe the landscapes preserved in national parks. What does this term really mean? *Webster's New Collegiate Dictionary* offers two definitions: (1) "characteristic of the earliest, or earlier, period or condition," and (2) "still pure, unaltered; unspoiled." Paleoecology gives us the opportunity to really test our ecosystems. In other words, you cannot really know if an ecosystem is "still pure, unaltered" unless you delve back into the ancient past to set a baseline. You have to probe those "earliest, or earlier," periods to find out the range of natural variability of the system. What were the extremes of temperature and precipitation? What species have been a part of the regional ecosystems? If we are to restore a given region to its pristine state, we have to know what that state should be. As mentioned previously, Indians had a part to play in

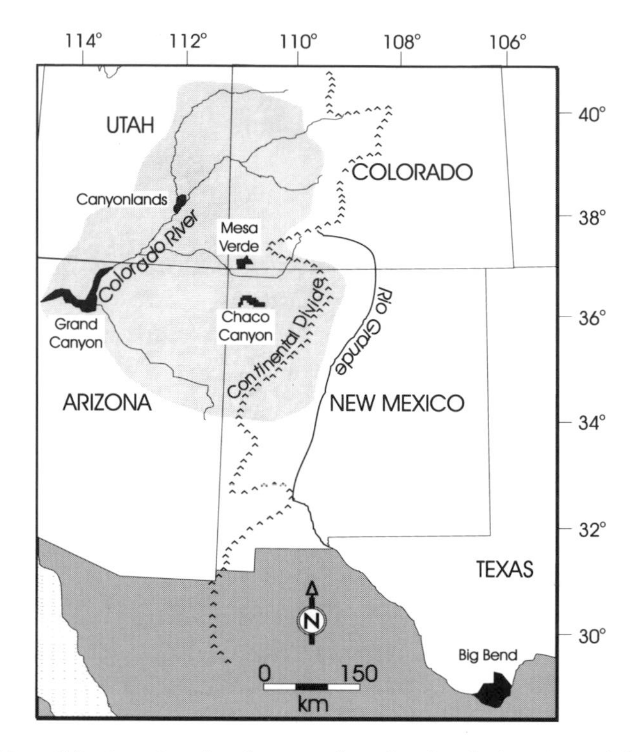

Figure I.2. Map of the American Southwest region, showing the locations of the national parks and monuments discussed in this book.

the drama of biotic change. To some degree, they had already "tamed" North America before Europeans got there. We have to be able to tease apart the results of Native Americans on the landscape from the results of environmental change. Those effects began almost 12,000 years ago, when the first ancient hunters arrived from Asia via the Bering Land Bridge. These are some of the challenges and controversies that will be explored in this book.

Let me issue one caution before we go any further. If you are reading this book, it is probably safe to assume that you already have some interest in fossils or archaeological artifacts. If you visit a national park and spot either fossils or artifacts, please do not touch them. The best thing to do is to mark the spot, or leave someone at the spot, and go find a park ranger. Tempting as it might be to pick up an arrowhead or a fossil bone and bring it back to a ranger station or visitor center, this would be the wrong thing to do. By removing a fossil or artifact from its original location, you would be destroying the evidence of exactly where it came from. This evidence is often vital in pinning down the age or the environmental or cultural context of the find. Fossils and artifacts that have been removed from the places where they were originally deposited cease to be useful to science; they become mere souvenirs, stripped of the context that gives them much of their meaning.

Speaking of souvenirs, it is, of course, illegal to remove fossils and archaeological materials from national parks (along with rocks, soils, plants, and animals, except fish caught with a park-issued license). Furthermore, all research done by qualified scientists in national parks is carried out only with the permission of the National Park Service. Science officers (resource managers) in the parks issue permits for specific projects at specific locations.

1

QUATERNARY FOSSILS

What Are They, and Where Are They Found?

Simply stated, Quaternary fossils are the remains of organisms that lived during the Quaternary Period. The Quaternary Period is the most recent of the geologic periods; it is the time interval covering the last 1.7 million years and is characterized by numerous glaciations. At the turn of the century, geologists thought that there had been just four major glaciations in North America during the Quaternary Period. Based on evidence from deep-sea cores and ice cores from Greenland and Antarctica, it now appears that North America experienced at least 17 Quaternary glaciations. The deep-sea cores contain microscopic fossils of marine organisms. The species composition of those fossil assemblages changes through time, reflecting the shifts from glacial (cold) to interglacial (warm) ocean temperatures. Ice core data show changes in the Earth's atmosphere, as revealed by minute bubbles of air that were trapped in the ice as it accumulated (as snow) over the last few hundred thousand years.

These multiple glacial intervals are lumped together into the **Pleistocene epoch,** which began with the onset of the first glaciation 1.7 million years ago and ended with the retreat of ice at the end of the last glaciation 10,000 years ago. The interval since the end of the last ice age is called the **Holocene epoch.** To put this interval in a human perspective, all of written history has taken place in the second half of the Holocene.

Humans have occupied the American Southwest during the last 12,000 years. This book focuses on events at the very end of the Pleistocene and the Holocene, a time interval short enough that few evolutionary changes in species are to be expected.

The last Pleistocene glaciation in North America, the **Wisconsin Glaciation,** was probably the most important force affecting the development of our modern ecosystems; it had an effect on every part of the continent, even if it did not cover the whole continent with ice. The American Southwest remained free of glacial ice, except in high mountain regions, but southwestern climates were radically altered during the glacial period. Retreat of the glacial ice in the north coincided with a shift to warmer, drier conditions in the Southwest, opening the way for recolonization of landscapes and a massive reshuffling of species in every region of North America.

Vertebrate Fossils

A number of types of Quaternary fossils are commonly studied by paleontologists. In the Southwest, these include the remains of large mammals, such as mammoths, giant sloths, and Pleistocene horses. Vertebrate fossils often dominate museum displays and popular literature. This notoriety is based on their high visibility and our own affinity with mammals. The skeleton of a mammoth has the capacity to capture our imagination, whereas most of us have a hard time getting excited about pollen grains, **diatoms,** or bits and pieces of insects. Although the fossils of large animals are fascinating and informative, they are actually quite rare in comparison with the fossils of smaller animals and plants.

It is sometimes difficult to make paleoclimatic interpretations based on the fossil remains of large animals, because many of them probably migrated across landscapes on a regular basis and thus were able to avoid undesirable climatic conditions. On the other hand, small mammals, such as rodents, shrews, and rabbits (not to mention fish, reptiles, and amphibians) offer tremendous opportunities for paleoenvironmental reconstructions, because they generally remained in their small home range year-round. For instance, fossils of the collared lemming, an inhabitant of arctic tundra, have been found in late Pleistocene deposits in the American Midwest. These data provide convincing evidence that the climate of central Iowa in the late Pleistocene was much cooler than it is now.

In addition, large mammals are far rarer than small mammals in any given ecosystem (for instance, there were many more mice than mammoths in North America during the Pleistocene). Consequently, a given fossil assemblage may produce thousands of bones (or pieces of bones) from voles, mice, and squirrels, but only a few bones of large mammals.

Although the vertebrate fossil record is important, scientists have gathered far more information from other kinds of fossils, most of which are much less glamorous; in fact, most are practically invisible to the naked eye. Among these are the pollen, stems, leaves, and fruits of plants; the **exoskeletons** of insects; the shells of snails and other mollusks; and the glassy skeletons of microscopic algae, called diatoms. These types of fossils are small, but they are much more abundant in sediments than the bones of ancient mammals. In fact, just a thimbleful of lake sediment or ancient soil on the floor of a cave may contain thousands of pollen grains, all wonderfully preserved down to the last detail of **microsculpture,** as viewed through a high-powered microscope.

By compiling the data gathered from all of the various types of fossils from a given time period in a study region, teams of scientists are able to piece together a picture of its plant and animal life. They can then use the paleontological data to reconstruct the history of climate change.

Fossils from Plants

Pollen

Palynology is the study of pollen. It is probably the most widely used tool in terrestrial Quaternary paleoecology. Many kinds of plants produce a superabundance of pollen each year. This is especially true of wind-pollinated plants, such as conifers (evergreens). A single lodgepole pine may produce as many as 21 billion pollen grains per year. Other plants, such as insect-pollinated species, may produce only a few thousand grains per year. These plants are usually underrepresented in the pollen rain, the fallout of pollen grains that showers down from the atmosphere. For instance, in the American Southwest, pollen from pine trees is often abundantly preserved in sediments, whereas pollen from junipers is much rarer, even in regions where junipers are quite abundant. Junipers simply do not produce as much pollen as pines.

Pollen has an extremely durable outer wall that resists decay. The pollen grains of many plant species are light enough to float on the wind and may travel hundreds or thousands of miles before landing. If they land in a lake, pond, or bog, they may be preserved for thousands of years.

Differences among the pollen of different species in a genus are often difficult or impossible to detect under a microscope. Because of this, pollen grains are most often identified only to the family or genus level rather than to the species (see Fig. 1.1 for an illustration of the taxonomic hierarchy). For instance, a pollen grain may be identified as coming from a pine tree, but it is not always possible to tell

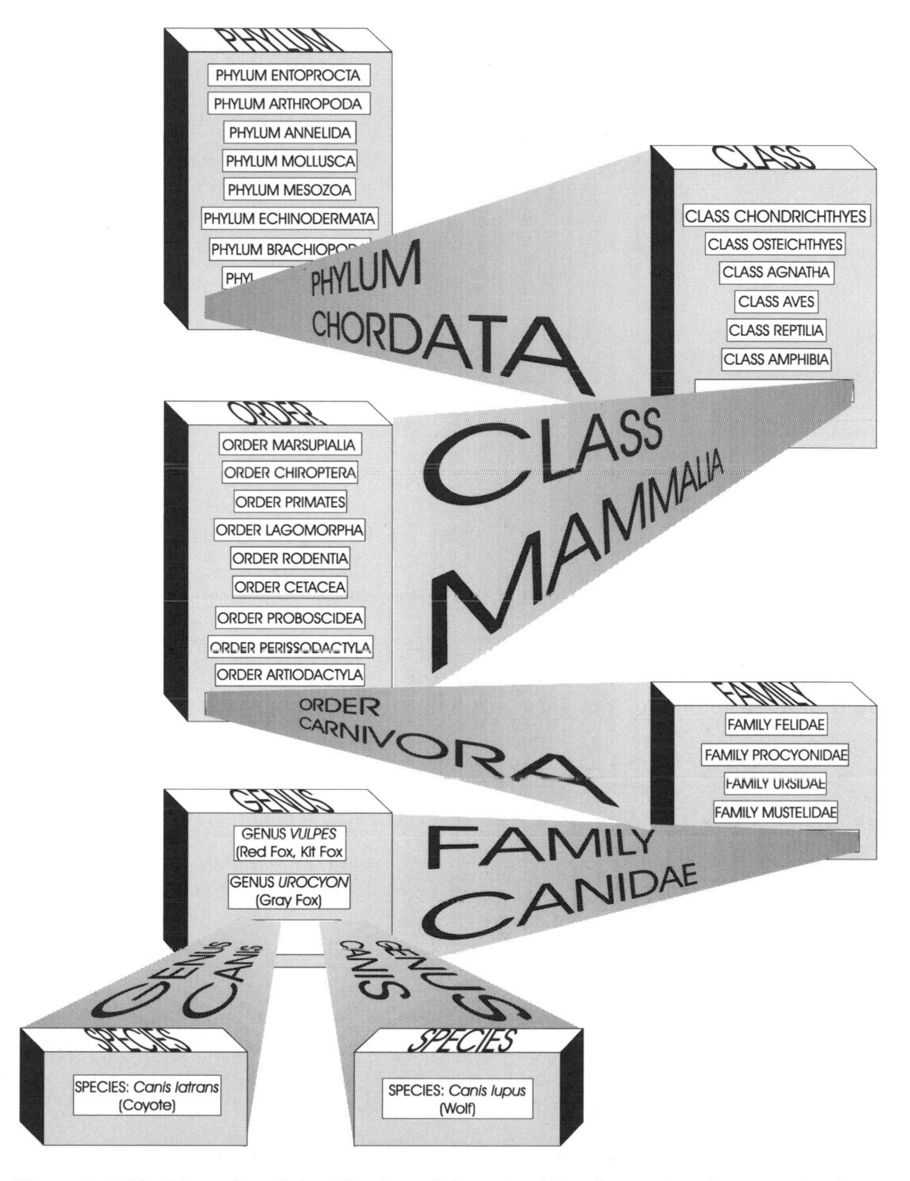

Figure 1.1. The hierarchy of classification of the animal kingdom, using the example of the wolf and coyote, two species in the genus *Canis,* which is one of several genera in the family Canidae (dogs, wolves, coyotes, and foxes). The canids represent one family in the order Carnivora (meat-eating mammals). The carnivores, in turn, are one order in the class Mammalia (mammals). The mammals are one class in the phylum Chordata (animals with spinal chords). There are many phyla in the kingdom Animalia (the animal kingdom).

from which species of pine (such as piñon pine, limber pine, or lodgepole pine) the pollen comes. This limits the accuracy of the interpretation of pollen data, since the ecological variability of a genus is necessarily greater than that of its component species.

Once the pollen grains are identified, the data are generally presented as diagrams showing pollen percentages of the total number of grains in sediment samples, plotted according to the age of the sample in a packrat midden or the depth of the sample in a column of lake sediment. Pollen diagrams from the same region and the same time interval are generally quite similar to each other but different from diagrams from other regions or times. Pollen diagrams are often divided into zones of similar pollen composition. The boundaries between zones mark transitions in regional vegetation. The proportion of pollen released into the environment depends on the number and type of plants present and therefore reflects the composition of regional vegetation. For instance, following the retreat of the Wisconsin Ice Sheets in North America, some regions were first colonized by low, herbaceous vegetation (grasses, sedges, and their relatives), similar to what is now found in the arctic tundra. As regional climates warmed and soils began to mature, the herbaceous vegetation gave way to coniferous forest, which in turn was invaded by hardwood trees, eventually producing the mixed deciduous-coniferous forests of today. In the Southwest, coniferous woodlands dominated by piñon pines and junipers gave way to **desert scrub** communities and desert grasslands, culminating in the modern vegetation within roughly the last thousand years. The timing of these transitions in plant communities is usually clearly shown in regional pollen diagrams.

If a given region has a mixed vegetation of piñon pine, juniper, and desert scrub plants, the pollen found in regional deposits may be dominated by pine grains, with proportionally fewer grains of juniper or desert scrub plants. It is therefore quite difficult to determine from the pollen assemblages if juniper or desert scrub plants were growing nearby. For instance, in Big Bend National Park, creosote bush dominates many lowland landscapes today, but its pollen is poorly preserved in regional packrat middens. On the Colorado Plateau, paleobotanists may have a hard time discriminating between piñon-juniper woodland and desert scrub communities based on their pollen "signatures." Plant macrofossils can help clarify the issue, however, since the remains (stems, fruits, leaves, or needles) of a wide variety of regional plants are gathered by packrats.

In the American Southwest, fossil pollen has been studied from both lake sediments and packrat middens. Airborne pollen dominates lake sediment pollen assemblages, but it is not the only source of pollen in packrat middens. Pollen is also brought into the den on flowers or foliage, sometimes in large quantities that may overwhelm the fallout from airborne grains.

Recently, palynologists have been developing statistical methods for reconstructing past environmental conditions, based on comparisons of the pollen "signature" of modern stands of vegetation with fossil pollen assemblages. By analyzing the climatic conditions under which the modern vegetation is growing and the proportions and amounts of pollen types in modern sediments from the various types of plant communities, they have been able to come up with fairly precise estimates of past climates. In addition, palynologists have used pollen diagrams to track the long-term movements of plant species and genera through the late Quaternary. Long-lived plants, such as most trees, do not spread rapidly across landscapes. For instance, junipers that dominated many landscapes in the Chihuahuan Desert region during the Wisconsin glacial interval took 500–1000 years to retreat upslope to higher elevations after the initial climatic warming at the beginning of the postglacial period. Such migrations in response to climate change may take centuries, but plants have undergone some remarkable shifts in distribution through the past few thousand years. This may seem like a long time to us, but it represents only a small fraction of the time that a given plant species has been in existence.

Plant Macrofossils

The macroscopic (visible to the naked eye) remains of plants, commonly preserved in Quaternary sediments, are called **macrofossils.** Woody plants produce a great deal of potential macrofossils throughout their life cycle. As might be expected, wood is very resistant to decay in poorly oxygenated, waterlogged sediments. In the Southwest, wood also preserves well in dry caves and packrat middens. Specialists in fossil wood identification are able to determine species of trees on the basis of analysis of cross-sections of stems. The annual growth rings in trees can provide both a history of local environments (for example, drought versus abundant precipitation, or episodes of forest fires or insect infestation) and a year-by-year chronology of those events. In archaeological studies, wooden timbers, beams, and poles used in buildings can be dated and analyzed for environmental reconstruction, based on the pattern of ring widths. Macrofossils from coniferous trees also include needles and cones, both of which can frequently be identified to the species level.

Nonwoody plants, such as grasses and herbs, also produce a wide variety of macrofossils, including stems, leaves, and fruits. Peat is essentially composed of layer upon layer of undecomposed plant leaves and stems. The two principal types of peats are moss peats (frequently dominated by sphagnum mosses) and sedge peats. Moss fragments can often be identified to the species level. Although peat accumulation is rare in the American Southwest, small bogs have been studied in various regions. These may persist for long periods of time, especially where they are fed by groundwater. Sphagnum mosses in bogs dominated many ice-free regions during the Pleistocene, but the American Southwest was generally too dry to support the growth of sphagnum.

In most temperate regions, plant macrofossils generally accumulate in **water-lain sediments,** and reflect only local conditions, since they come from plants growing in the **catchment basin.** In order to develop a regional reconstruction of past vegetation based on macrofossils, it is necessary to study multiple sites and piece together the data from each site into a synthesis of information. This is likewise true of macrofossils found in packrat middens.

Packrat Middens

Packrat middens are the most important repository of both pollen and plant macrofossils in the American Southwest. The discovery of packrat middens as a source of fossil plants and small animals is undoubtedly one of the most important advances in Southwestern paleontology. These middens are caches of objects—including edible plants, cactus spines, insect and vertebrate remains, small pebbles, and feces—brought to the den site for a variety of reasons, including consumption as food, satisfaction of curiosity, and protection of the den. Once in the den, they are cemented into tan, brown, or black tarry masses by packrat urine. When dry, a midden hardens into a durable, resistant mass that can preserve a paleoecological record for thousands of years in a dry rockshelter.

Packrats, or woodrats, are the native North American rats of the genus *Neotoma.* There are 21 species of packrats; their range today extends from arctic Canada to Nicaragua. They are a very successful group, inhabiting a variety of habitats. Packrats are not strict **herbivores,** but plants make up by far the most important part of their diets. Packrats are den builders and essentially solitary. Regardless of size, each den is occupied by one individual, or by a female and her young. These dens range from piles of sticks under shrubs or trees in lowland regions to more elaborate shelters in caves and rockshelters on hillsides. In the American Southwest, the dens often include tightly packed bundles of cactus stems (for instance, cholla), which keep out predators. The walls of many dens consist of finely shredded plant material, which serves as insulation. Beside bedding material and food plants, packrats also accumulate other objects from the surrounding landscape, including bones, pieces of dung from large animals, feathers, or other objects that capture their attention (dens have been found to contain false teeth, silverware, cigarette lighters, aluminum foil, and prehistoric potsherds). These objects, plus packrat feces, accumulate in middens that are cemented together by layer upon layer of crystallized packrat urine, or amberat. When protected from direct rainfall and high humidity, amberat is extremely resistant to weathering and decomposition. It locks biological specimens into a desiccated time capsule. In the American Southwest and northern Mexico, many packrat middens have remained intact for more than 45,000 years (i.e., beyond the limit of the radiocarbon dating method [see Chapter 3]).

Packrat middens continue to accumulate as long as a den is occupied. In rockshelters, caves, or crevices, favorable denning sites are inhabited by generation after generation of packrats, so their middens build up for hundreds and perhaps thousands of years. Packrat middens are an unusual source of fossils. Most Quaternary fossils accumulated in lakes, ponds, and streams because they were washed or blown into them, but packrat middens were manufactured by packrats, so only those objects accumulated by the rats end up in their middens. They eat a wide variety of plants, and unconsumed pieces of these plants form the bulk of the material in midden records.

The urination perch is the site where the rat urinates and defecates. Rats urinate frequently, keeping the surface of the perch moist and sticky. This sticky surface acts as a trap for airborne pollen. Dung beetles and other dung-feeding insects are attracted to it, and other arthropods also get trapped in the sticky matrix of feces and urine. The bones of mammals and reptiles are often found in packrat middens, as are the bony plates and scales of reptiles. Owls and other birds of prey sometimes nest in caves and rockshelters. They leave behind the undigested fur and bones of their prey (mostly small rodents), and these may also end up in middens. Sometimes packrats pick these up from the cave floor and add them to the accumulation of midden debris. The remains of other cave and rockshelter inhabitants, such as mice, may also end up in packrat middens, either by accident or when the packrat adds these remains to its den as part of its interior decorating scheme. Geologist Robert Thompson of the U.S. Geological Survey found a packrat den in Nevada that was "decorated" with wall-to-wall scorpion tails. Apparently human beings are not the only hunters who like to display their trophies!

Paleoecologists working in the American Southwest owe a large debt of gratitude to packrats. Without the fossil record preserved in packrat middens, we would know far less about the biological history of this arid region. Yet a great deal more remains to be learned as this new technique for probing the late Quaternary is refined.

Plant macrofossils in packrat middens provide the most useful information on the history of regional vegetation. Packrats forage for a wide variety of plant foods, and midden plant assemblages may contain dozens of species. Whereas the pollen record leaves doubts about the proximity of the source plant to the fossil site, the plant macrofossil record is relatively unequivocal on this subject. Modern studies of packrat foraging behavior show that the rats rarely travel more than a few

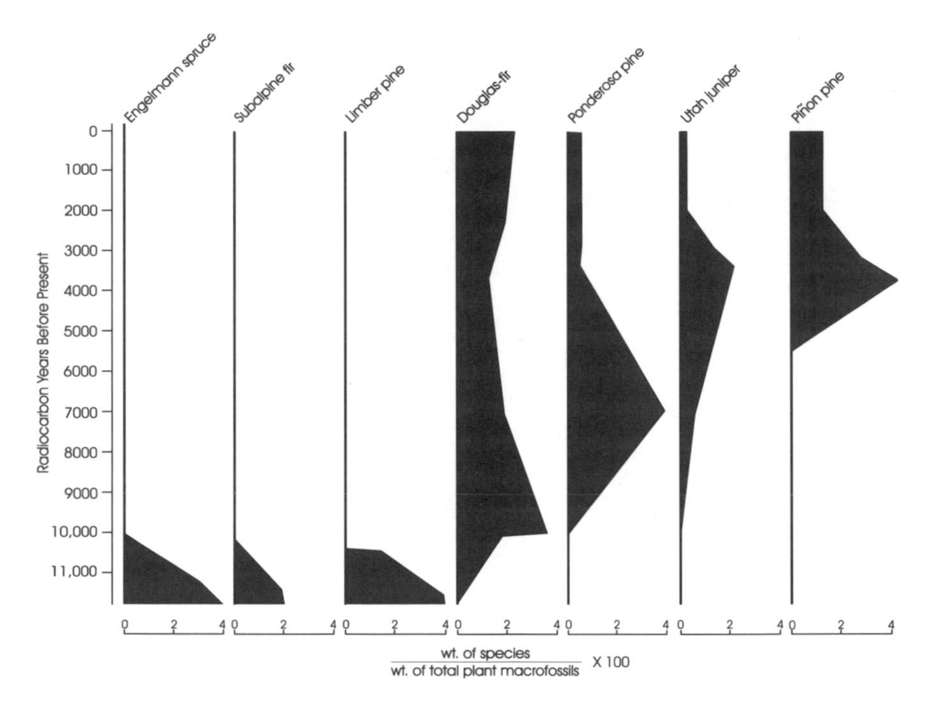

Figure 1.2. Generalized plant macrofossil diagram for packrat midden assemblages from sites in the American Southwest during the last 12,000 years. The diagram reflects changes in the relative abundance of each species in the assemblages through time. (Modified from Betancourt et al., 1990.)

hundred meters from their dens in search of food. Therefore, the plant species found in a packrat midden almost certainly grew within a stone's throw of the cave or rockshelter when the midden was deposited. Moreover, good denning sites are rarely left unoccupied for long. In a study I did with mammalogist James Halfpenny, a packrat died that had been living in a nicely protected rockshelter with abundant food resources close by. Within twelve hours of its death, a new packrat had moved in. This packrat behavior translates into more or less continuous occupation of some rockshelters and shallow caves, which in turn leads to the nearly continuous accumulation of middens. This kind of uninterrupted fossil record, chock full of the remains of plants that all lived nearby, allows paleobotanists to trace the shifting ranges of plant species through time with great temporal accuracy and geographic precision. Plant macrofossil assemblages may be tedious to examine in layer after layer of midden at site after site throughout a region, but once the work is done, the paleobotanist may develop a blow-by-blow reconstruction of regional vegetation that is unrivaled anywhere else in the world (Fig. 1.2).

Tree Rings

In the American Southwest, the study of tree rings has been one of the principal methods used to reconstruct the history of environmental change during the late Holocene.The study of climate change based on tree rings is called *dendroclimatology;* it is carried out in conjunction with tree ring dating, or *dendrochronology,* which is discussed in Chapter 3. These research tools were originally developed for work on the Colorado Plateau. Several species of conifers that grow in the southwest are very long lived, especially the whitebark and bristlecone pines. Bristlecone pine is the organism with the longest known lifespan: some individual trees in the White Mountains of California have been alive for several thousand years. Many trees form two rings per year. One ring, made up of large cells, is added during the growing season of spring and summer; the other ring, made up of small cells, represents the more or less dormant period of fall and winter. The types of cells laid down in these two sets of rings are usually easily distinguished from one another. Differences in the size of growth rings laid down during the life of a tree are used to interpret changing environmental conditions: years of abundant warmth and moisture yield broad growth rings, whereas years of drought or cold summers yield narrower rings.

However, other factors—including the age, the height, and species of the tree—also affect the size of growth rings. Dendrochronologists compensate for these factors by standardizing the data from different trees in a given region, creating *ring-width indexes.* The standardardized indexes of individual trees are averaged to obtain the mean (averaged) chronology for a sample site. After the ring widths for a given tree sample have been measured, an average ring width is obtained, then deviations from that average are plotted. Single-year averages are not as useful as averages taken over 5–10 years of tree growth, because ring width in a single year may be affected by moisture conditions over the previous three years. Tree ring index chronologies are applied to the trees growing in a single region, where, presumably, the climatic influences on tree growth have been fairly uniform.

Recently, tree ring researchers have developed other types of studies to examine different aspects of forest paleoecology. For instance, Tom Swetnam and Julio Betancourt have published a report on the frequency of fires in the Southwest, based on analyses of fire scars that show up in tree rings. They compiled fire scar records from tree rings in ponderosa pine, piñon pine, and Douglas-fir stands in New Mexico and Arizona, then compared fire frequency patterns in the tree rings to the pattern of El Niño, an oscillation in tropical Pacific ocean currents that strongly affects weather patterns from the northwest coast of South America through the Four Corners region of the United States and beyond.

Insect Fossils

Insect fossil studies began in earnest during the 1960s and have become one of the most important sources of terrestrial data on past environments. These studies have, for the most part, employed fossil insects as **proxy data,** that is, as indirect evidence for past environmental conditions.

Beetles are the largest order of insects. They have been the main insect group studied from Quaternary sediments, and in fact they are the most diverse group of organisms on Earth, with more than one million species known to science (that's more than all of the flowering plants combined). In addition, their exoskeletons, reinforced with **chitin,** are extremely robust and are commonly preserved in large numbers in lake sediments, peats, and some other types of deposits. In most cases, beetles have quite specialized habitats that apparently have not changed appreciably during the Quaternary. This characteristic makes them excellent environmental indicators. The exoskeletons of beetles and some other insects are covered with exquisite microsculpture, enabling paleontologists to identify fossil exoskeletons to the species level in at least half of all preserved specimens, even though insect exoskeletons are most often broken up into the individual plates in fossil specimens.

Beetles are very quick to colonize a region when suitable habitats become available. They often respond more quickly than long-lived plants such as trees and shrubs, which, until recently, were relied upon almost exclusively as indicators of environmental change on land. Like plant macrofossils, insect fossils are generally deposited in the catchment basin in which the specimens lived. Thus they provide a record of local conditions, in contrast to pollen, which can be carried many miles on winds and often gives a more regional "signal."

Studies of insect fossils in two-million-year-old deposits from the high arctic have failed to show any significant evidence of either species evolution or extinction. Beetle species have apparently remained constant for as many as several million generations. Presumably, their ecological requirements have also not changed during this interval, as attested to by the constancy of ecologically compatible species assemblages through time. Although the datable fossil insect record from the American Southwest does not extend back more than about 50,000 years (the limit of radiocarbon dating), beetle species in this region appear to have been stable at least that long, and probably much longer.

In the American Southwest, nearly all fossil insect research has been based on remains found in packrat middens. Most insects that are preserved in middens represent species that live outside caves and rockshelters (Fig. 1.3). These are mostly predators and scavengers that come into the packrat's den to make use of food resources (dung beetles, nest parasites, scavenging insects that remain in and

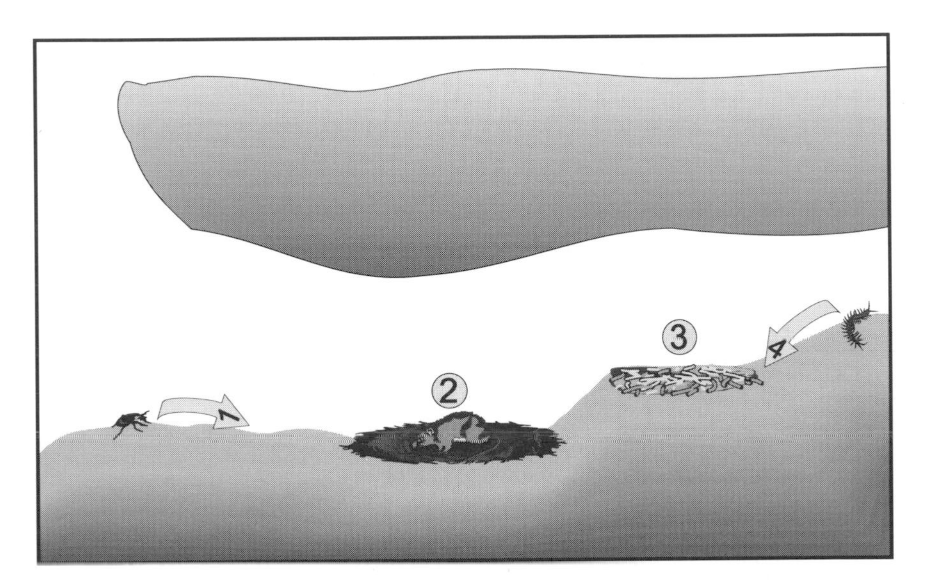

Figure 1.3. Sources of insects and other arthropods preserved in packrat middens. 1, Insects that live outside the cave and come in to hunt, scavenge, or escape bad weather; 2, packrat parasites and nest inhabitants; 3, dung beetles and detritus feeders; 4, cave dwellers.

around the nest). In winter, insects are apparently attracted to packrat dens because of their warmth. In regions outside the American Southwest, insect fossils are generally extracted from organic-rich lake or pond sediments or peats. Ancient stream flotsam, deposited in **fluvial sediments** and later exposed along stream banks, is often a rich source of insect fossils.

Generally, insect fossil lists include many species, but just a few individuals of each. Paleoclimatic reconstructions are usually made on the basis of the climatic conditions in the regions where the species in a given assemblage can be found living together today, that is, the climate of the regions where their modern distributions overlap.

Aquatic Organisms

Ostracods and other organisms can provide information on ancient lake conditions. This, in turn, can often be tied directly to climate, because the size and water quality of lakes are controlled largely by the balance between evaporation, precipitation, and local ground-water conditions. The fossil organisms in lakes reflect

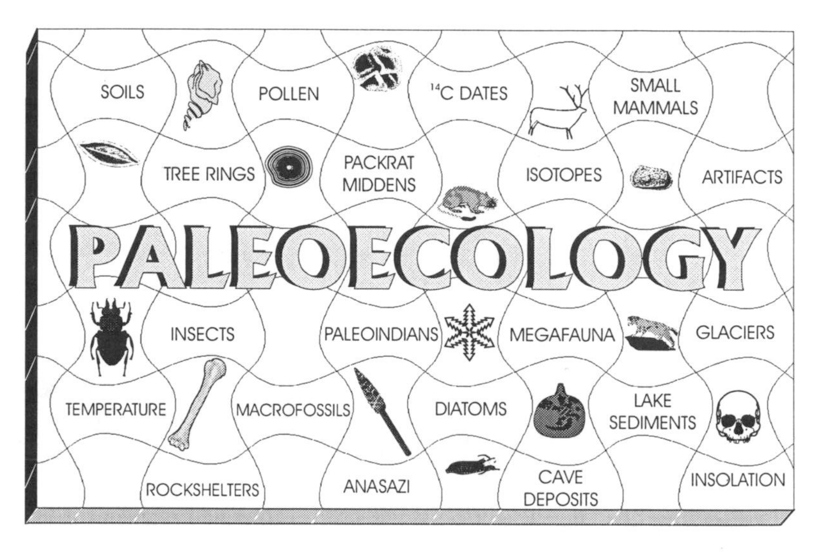

Figure 1.4. In some ways, paleoecological reconstruction resembles the piecing together of a jigsaw puzzle.

changes in regional climate, as those changes affect lake water temperature, salinity, alkalinity, and other factors. However, sometimes local factors, such as the presence of springs or **alkaline sediments,** may overwhelm the regional climatic signal.

Ostracods are tiny, bivalved crustaceans that live in fresh or salt water. Their shells, or carapaces, are reinforced with calcite (crystalline calcium carbonate), which dissolves in nonalkaline waters. Their fossil shells therefore survive only in nonacidic sediments, such as **marl.** The abundance and diversity of ostracods in a given lake are usually dependent on salinity, amount of oxygen, acidity–alkalinity, water depth, and food availability. Some are found only in lakes; others prefer ponds; still others live in running water. The alkaline sediments that accumulate in the standing water of desert regions often contain ostracods.

A host of other "critters" in lakes and ponds leave behind fossils that can serve in paleoenvironmental reconstructions. Among these are freshwater sponges, water fleas, and snails (both terrestrial and aquatic). Snails are particularly important in the fossil record of the American Southwest, because their shells preserve well in alkaline sediments.

A variety of these aquatic plants and animals can be found in abundance in water-lain sediments. An equally varied collection of fossils is often found in packrat middens, including plant macrofossils and pollen, insects and other arthropods (including spiders, scorpions, and millipedes), the bones and fur of mammals, and the bones and scales of reptiles. It may seem redundant to study more than one or two types of fossils, since many of them provide overlapping information on past environments. However, paleoecologists don't look at it that way. They consider each piece of fossil data to be relevant, because each fossil group usually has some unique data to contribute, and the combined information always makes for a better, more sharply defined picture of past environments. If nothing else, different data sets serve as corroborative evidence, confirming the information provided by some of the more commonly studied groups.

As we have seen, paleoecology is basically detective work, with fossils as witnesses to past environments. The fossils have a lot to say if you can understand their language. The process of paleoenvironmental reconstruction is a lot like fitting together pieces of a jigsaw puzzle. The more pieces you have, the more successful you'll be (Fig. 1.4).

Suggested Reading

Birks, H. J. B., and Birks, H. H. 1980. *Quaternary Palaeoecology.* London: Edward Arnold. 289 pp.

Eicher, D. L. 1976. *Geologic Time.* Englewood Cliffs, New Jersey: Prentice Hall. 150 pp.

Elias, S. A. 1994. *Quaternary Insects and Their Environments.* Washington, D.C.: Smithsonian Institution Press. 284 pp.

Spaulding, W. G., Betancourt, J. L., Croft, L. K., and Cole, K. L. 1990. Packrat middens: Their composition and methods of analysis. In Betancourt, J. L., Van Devender, T. R., and Martin, P. S. (eds.), *Packrat Middens: The Last 40,000 Years of Biotic Change.* Tucson: University of Arizona Press, pp. 59–84.

Swetnam, T. W., and Betancourt, J. L. 1990. Fire–southern oscillation relations in the southwestern United States. *Science* 249:1017–1020.

Warner, B. G. (ed.). 1990. *Methods in Quaternary Ecology.* Geoscience Canada Reprint Series No. 5. St. John's, Newfoundland: Geological Association of Canada. 170 pp.

2

THE REPOSITORIES OF ECOLOGICAL HISTORY

Where Are Ice-Age Fossils Found?

When I mention to people that I work on insect fossils, they often assume that I have to hack away at outcrops of shale, looking for faint impressions of fossils in the stone. In fact, Quaternary fossils are most often found in mud, not in stone. That is because most Quaternary sediments are **unconsolidated,** that is, they are loose soils, sands, clays, and other matter that have not yet turned to stone (given a few million years more under pressure and/or great heat, they may). Quaternary fossils can also be found in peat moss (the kind used for gardening). Peat is composed of waterlogged plants that accumulate in bogs because of extremely slow decomposition. Given a few million years under the right conditions, peat turns to coal.

Pre-Quaternary fossils in bedrock generally fall into one of two categories: they are either mineral replacements of the original organic matter or impressions of ancient plants and animals. In both cases, the original organic matter has long since decomposed. However, most Quaternary fossils are the actual remains of plants and animals that have been preserved before decomposition set in. As mentioned previously, this type of preservation occurs in peat bogs. It also takes place at the bottom of lakes and ponds, where the sediments are low in or without oxygen (oxygen supports aerobic bacteria that decompose and destroy the structure of organisms). In very dry environments, such as desert caves and rockshelters, the remains of plants and animals gradually dry out and may become "mummified" fossils that can last tens of thousands of years. In packrat middens, as we saw in Chapter 1, plant and animal remains become part of a matrix of rat feces and urine that dries into rock-hard amberat. This hard, dry matrix preserves fossils remarkably well, and it readily dissolves in water, making it relatively easy to release the fossils from their rather unusual tomb. Middens survive for many thousands of years in arid regions, but only in localities such as shallow caves and rockshelters, where they are protected from moisture.

The dried packrat urine that is preserved in middens has recently provided yet another source of paleoenvironmental information. Chemist Pankal Sharma studied late Pleistocene and Holocene packrat urine from a midden from western Nevada in order to reconstruct the history of cosmic ray bombardment of the Earth. Cosmic rays that strike the Earth's atmosphere generate a radioactive isotope of chlorine (^{36}Cl). This isotope is taken up by plants, then passed to plant-eating animals, such as the packrat. Most animals do not preserve their urine for posterity, but packrat urine becomes incorporated into the amberat matrix, preserving a record of radioactive chlorine isotopes that serve as a proxy for levels of cosmic ray bombardment. Midden samples dated at 21,000 years before present (yr B.P). yielded packrat urine containing increased ^{36}Cl. This translates into a cosmic ray influx that was 41% higher than that seen today. Packrat urine chlorine from a sample dated at 12,000 yr B.P. showed a 28% higher influx of cosmic rays than that observed today.

Another new technique being developed using packrat midden samples is the reconstruction of past carbon dioxide (CO_2) levels in the atmosphere, based on evidence from plant macrofossils. Julio Betancourt and Pete Van de Water have been studying plant remains, looking at the numbers of pores in plant leaves where atmospheric gases are taken in. Plants breathe through these pores, called *stomata.* Modern studies in controlled atmospheric chambers show that the number of stomata plants develop are directly related to the concentration of CO_2 in the air. When CO_2 concentrations are higher, the plants need fewer stomata. These stomata are perfectly preserved in many plant macrofossils in middens, and Betancourt and Van de Water have confirmed findings previously made in ancient atmospheric gases trapped in ice cores from Antarctica. Both records (plant stomata and ice core gases) show that the CO_2 concentration in the atmosphere was substantially higher at 11,000 yr B.P. Increased CO_2 concentrations in today's atmosphere are thought to be one of the main contributors to "greenhouse gas" warming. Greenhouse gases act to trap the warmth of the Earth and keep it from radiating back out into space, much as the glass in a greenhouse keeps the warmth trapped inside the greenhouse. Increased CO_2 levels at 11,000 yr B.P. may have been one of the principal causes of the climatic warming that brought the last glaciation to an end.

Quaternary fossils are found in a wide variety of sediments. We will consider only terrestrial and freshwater sediments in this chapter. These materials are quite familiar to everyone. They are the soil in your garden, the smelly mud at the bottom of a pond, and the dust blowing into your house during a windstorm.

Types of Sediments Containing Fossils

In the American Southwest, mineral sediments, such as gravel, sand, silt, and clay, accumulate in three main types of deposits: eolian, alluvial, and lacustrine.

Eolian deposits are wind-blown silts or sands. Loess is a deposit of relatively uniform, fine sediment (mostly silt) that was transported to the deposition site by wind. Most loess deposits are relatively inorganic and contain few fossils. Loess covers large regions of North America, including Alaska, Washington, and Nebraska, in depths of up to several tens of meters.

Sand dunes are another type of eolian deposit. Although dunes do not mantle as much of the American Southwest as is depicted in some westerns, they can be an important repository of fossils. For instance, anthropologists at the University of California, Riverside, found dried, prehistoric human feces preserved in sand dunes near Palm Springs, California. The Cattleguard Paleoindian archaeological site in the San Luis Valley of Colorado, excavated by Peggy Jodry and Dennis Stanford of the Smithsonian Institution, is situated in and around sand dunes that probably helped preserve the stream sediments that contain butchered bones and stone tools. This site is very close to the Great Sand Dunes National Monument, and it is likely that large dunes covered the site many times in the last 10,000 years.

Alluvial or stream deposits are created by moving water; they include gravels, sands, and silts. Stream deposits often include pockets of organic debris, which accumulated in backwaters, pools, or low-energy side channels. These deposits can be a treasure trove of fossils because they consist of stream flotsam, which may accumulate in large quantities over a short time. When river channels shift, they sometimes abandon a side channel, which then becomes an oxbow lake. Once an oxbow lake forms, it rapidly fills in with vegetation and eventually becomes dry land. In the process, however, the water-lain organic detritus in the oxbow becomes part of the fossil record.

In the arid Southwest, alluvial deposits are often exposed in eroded stream channels called **arroyos.** These deposits have yielded well-preserved pollen and other fossils. For instance, palynologist Steve Hall was able to reconstruct the history of vegetation at Chaco Canyon National Monument by studying the sequence of pollen deposited in alluvial deposits in and around Chaco Wash. He sampled the pollen-bearing sediments from arroyos.

Organic deposits are also contained in buried soils. These occasionally include usable fossils, notably pollen. However, most organic matter in soils is more or less decomposed because of exposure to oxygen, so fossil preservation is generally not as good as in water-lain sediments. The other problem with buried soils is that the process of soil development does little to concentrate fossils, so they are few and far between.

Water tends to concentrate organic detritus (such as stream flotsam and organic debris in lakes) into recognizable layers that eventually produce more specimens per cubic centimeter than neighboring upland soils. Therefore, the principal types of sediments containing abundant fossils are those that are laid down in water. Water in a basin, such as a pond or lake, may collect large amounts of plant and animal matter from its watershed. Dead insects, leaves, seeds, and twigs are carried to ponds and lakes by streams. Insects, pollen, and small seeds are blown in by the wind or washed in by streams. Some animals (both large and small) simply fall into the water and drown or are washed downstream after they die (Fig. 2.1).

Not all lake sediments are alike. Sediments rich in calcium carbonate, known as marls, preserve fossil bones and the shells of mollusks and ostracods better than sediments that are more **acidic.** In some lakes, the layers of sediments laid down each year are poorly defined, whereas in others there are clearly marked couplets of sediment (winter and summer sediment layers) for each year. These annual layers are called *varves.* Varved sediments allow researchers to count back year by year in the fossil record of a lake, leading to incredible accuracy in the dating of events.

In addition to fossil preservation, a great deal of information about past environments can be gleaned from the study of the physical and chemical properties of lake sediments. The size of the particles in the sediment (gravel, sand, silt, and clay) reveals whether they were deposited in a low-energy environment (in the lake, or from a slow-moving stream) or a high-energy environment (from a river, a flooding creek, or a beach gravel redeposited after a severe storm). Most deposition takes place in low-energy environments, whereas erosion takes place in high-energy environments.

In the American Southwest, the last glaciation brought increased precipitation and cooler temperatures that slowed evaporation of water from the land surface. The net result of these climatic conditions was the creation of several very large lakes, called **pluvial lakes.** These lakes filled large basins in Utah and Nevada (Fig. 2.2). At its highest stage, Lake Bonneville covered much of western and central Utah, with a surface area of about 51,300 km^2 (19,800 square miles). Its greatest depth was more than 370 m (1200 ft). The Bonneville salt flats are the hardened, dried-out crust of the ancient lake bed. All that remains of it today is the Great Salt Lake. Although the size of this lake is impressive, it contains only a small

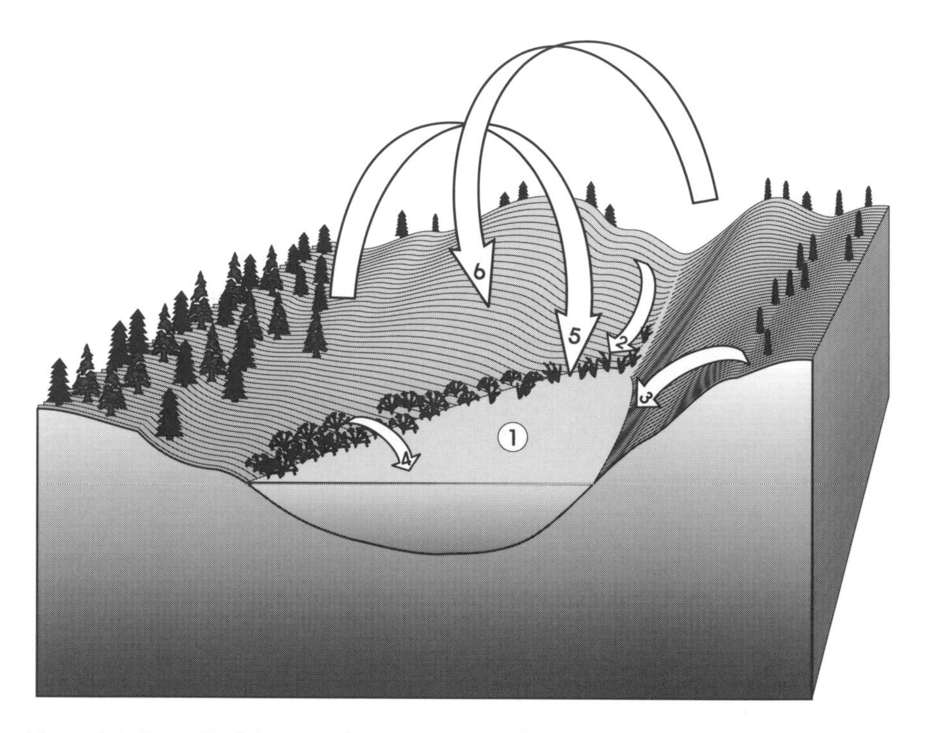

Figure 2.1. Generalized drawing of a mountain watershed, showing sources of potential fossils that may be deposited in a lake. 1, Plants and animals that live in the lake; 2, plants and animals carried into the lake by a stream (including stream-dwellers and organisms that fall into the stream); 3, plants and animals transported to the lake by slope wash, erosion, or solifluction; 4, lakeshore plants and animals that fall into the water; 5, microfossils (mostly pollen and spores) from nearby plant communities, transported to the lake by wind; 6, microfossils (mostly pollen and spores) from distant plant communities, transported to the lake by wind.

fraction of the water held by its pluvial predecessor in the late Pleistocene. Pluvial Lake Lahontan covered much of the basin east of the Sierra Nevada range in Nevada. At its maximum, this lake had a surface area of 22,300 km^2 (8600 square miles), and it had a maximum depth of 276 m (890 ft). The huge quantities of water in these lakes altered regional climates significantly, just as the Great Lakes alter the climate of the upper Midwest today.

Cave Deposits

In a generic sense, the term *cave* covers subterranean caverns of all sizes; the term *rockshelter* is more appropriate for most shallow caves inhabited by prehistoric

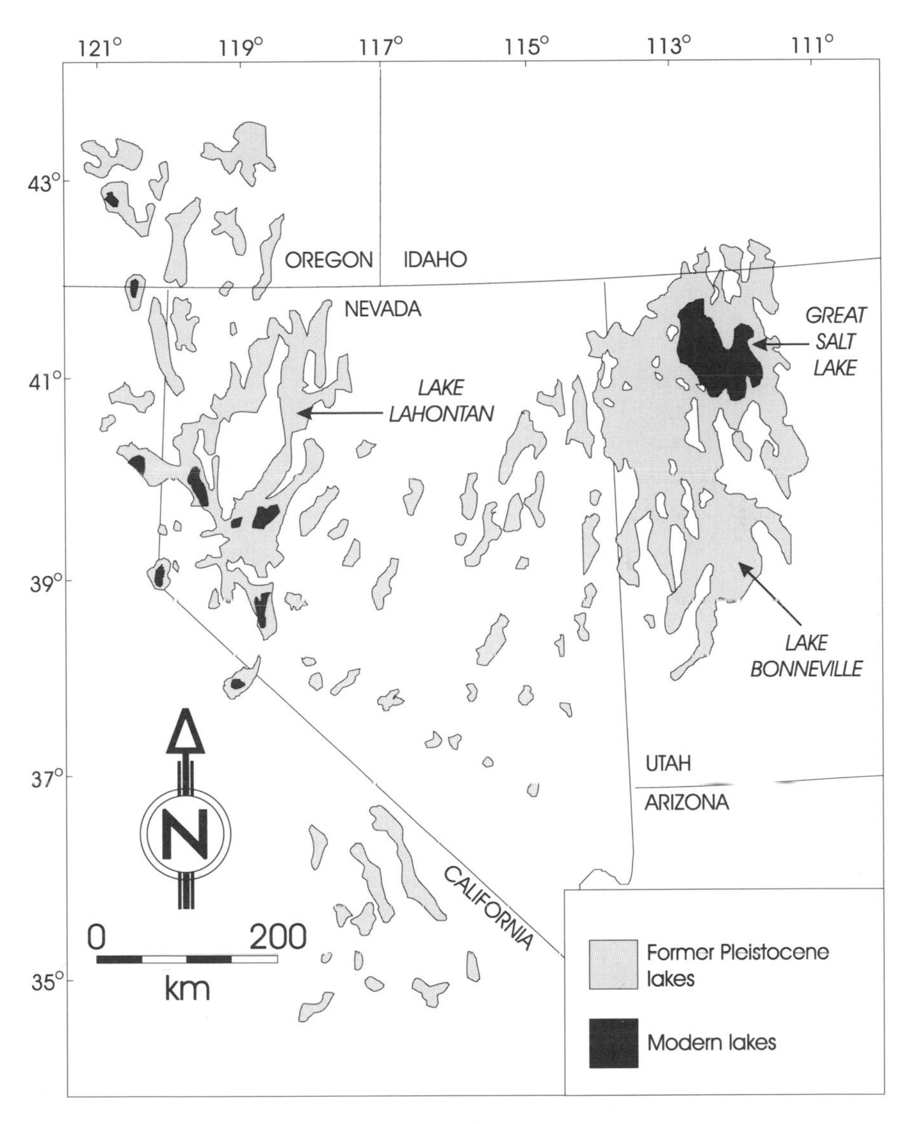

Figure 2.2. Map of the Great Basin region, showing former Pleistocene lakes and modern remnants. (After Benson and Thompson, 1987.)

animals and people in the American Southwest and elsewhere. Rockshelters are shallow enough to allow considerable sunlight and air circulation, even at their deepest point, whereas caves may extend many miles underground and include regions of complete darkness and little exchange of air with the outside world. I use the term *cave* here in the more generic sense.

For our purposes, cave deposits are most important as a source of vertebrate fossils and packrat middens. Much, if not most, of the fossil evidence described in the upcoming chapters on southwestern parks came from cave deposits. Caves are essentially closed systems, limited to walls, floor, and roof. They have a beginning (cave formation) and an end (collapse of the cave roof or infilling by outside sediments). Such temporal boundaries mean that the sediments (and associated fossils) that collect in a cave represent a discrete time interval; some caves persist for hundreds of thousands of years, others for a few tens of thousands of years. This is in contrast to many other types of sediment deposition, such as that in large lakes or rivers, where sediments may accumulate for tens or hundreds of thousands of years. Accumulation of sediments in a cave begins when an opening forms to the outside world. Most caves that have produced abundant vertebrate bones are small, shallow caves that are close enough to the surface to have substantial contact with the world outside the subterranean cavern.

There are five sources of bones in cave deposits (Fig. 2.3):

1. Animals carried in by predators (including both ancient and modern people).
2. Fecal or regurgitated pellets of animals or birds frequenting or living in the cave.
3. Predators or scavengers that are attracted to carrion, then become trapped after entering the cave.
4. Animals that fall into the cave through cracks or other openings from the surface.
5. Cave-dwelling animals.

Other than specialized cave dwellers, most animals will live only in caves that afford easy access to the outside world. Pleistocene caves provided shelter for numerous animals, including humans. When they died in caves, their remains were often preserved in cave-floor sediments. However, cave researchers have concluded that most bones in cave deposits are from animals that accidentally fell in, rather than from actual cave dwellers. A good example of this kind of cave is Natural Trap Cave, Wyoming. It has a small opening at the top, which widens to form oversteepened walls inside. Such caves often contain deposits with an abundance of carnivores (mountain lions, wolves, and other meat-eating mammals). It is thought that the carnivores are attracted to the smell of carrion emanating from the opening at the top of the cave and then enter (or fall into) the cave themselves and are unable to get out.

Predators, including birds of prey, often carry prey animals into the open mouths of caves. The bones of the prey are then deposited in the cave as refuse left over from the predator's meals, in regurgitated pellets, or in the predators' scat (feces).

Figure 2.3. Generalized section of a cave, showing potential sources of fossil bones: 1, predators that carry prey animals into the cave; 2, owl and other raptor pellets; 3, predators attracted by carrion that fall into the cave through large vents; 4, animals that fall into the cave through smaller vents; 5, animals that live in the cave. (Modified from Andrews, 1990.)

Taphonomy: How Fossils Become Preserved

The process by which a living organism becomes preserved in the fossil record is called **taphonomy.** Paleontologists have devoted entire careers to the study of taphonomy, or "the laws of burial." Figures 2.1 and 2.4 present brief summaries of taphonomic processes for a lake or pond in a catchment basin. Although the figures are oversimplifications of a very complex set of events, they shows the basic sources of potential fossil material on a living landscape and the general processes that lead to the preservation of plants and animals in sediments.

Most organisms that end up in the bottoms of lakes and ponds spent their lives either in the water or in close proximity to it. In fact, it makes intuitive sense that more aquatic and riparian (shore-dwelling) organisms are preserved in water-lain sediments than upland species, simply because the odds are smaller that the

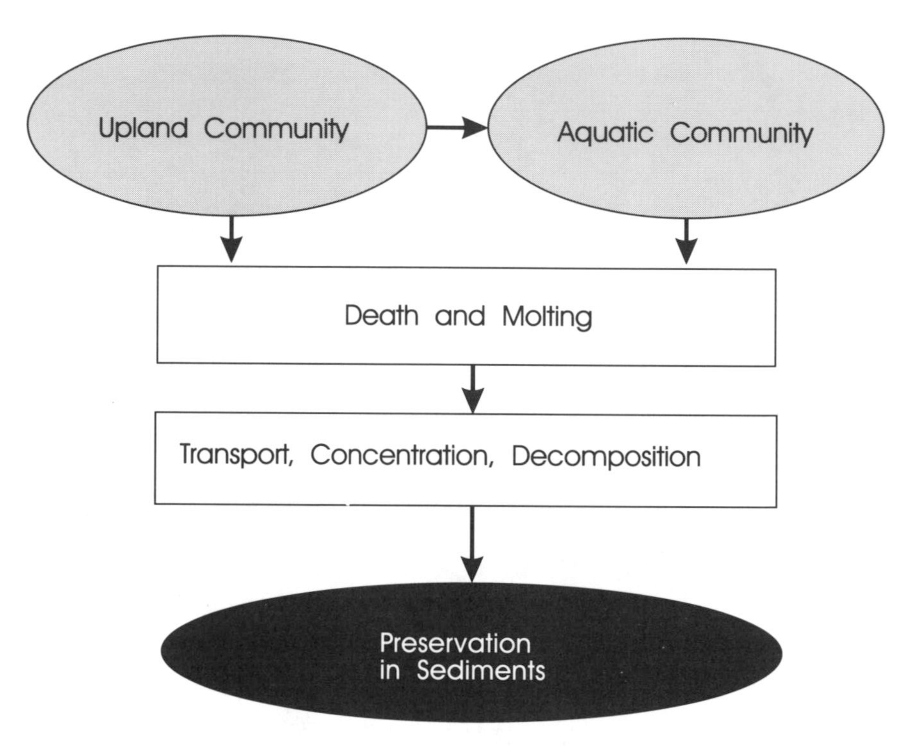

Figure 2.4. Summary of the taphonomy of fossils in a catchment basin.

remains of an upland creature will make their way into the lake. The remains of many upland creatures, both plants and animals, decompose on the surface and are not preserved as fossils. Even the boniest, most hard-bodied creatures will decompose if left to rot on a hillside. Some large upland animals die near or in the water or are washed downslope by rains, so their carcasses end up in water-lain sediments, but on the whole we have a better understanding of what life was like in or near the water and a poorer knowledge of ancient life on dry hillsides.

Another confounding factor is that some aquatic invertebrates (such as caddisfly larvae and ostracods) go through several stages of development before reaching maturity. With each new stage, they shed their old exoskeleton or shell. This molting process creates several sets of potential fossils for each individual.

Microfossils, including many types of pollen grains, float through the air and travel quite readily for many miles. This characteristic ensures that the pollen that rains out of the sky is representative of a broad region, including uplands and lowlands. Wind-pollinated plants, such as conifers, are overrepresented in the pollen records of lakes, because even though aquatic plants are assured of pollen

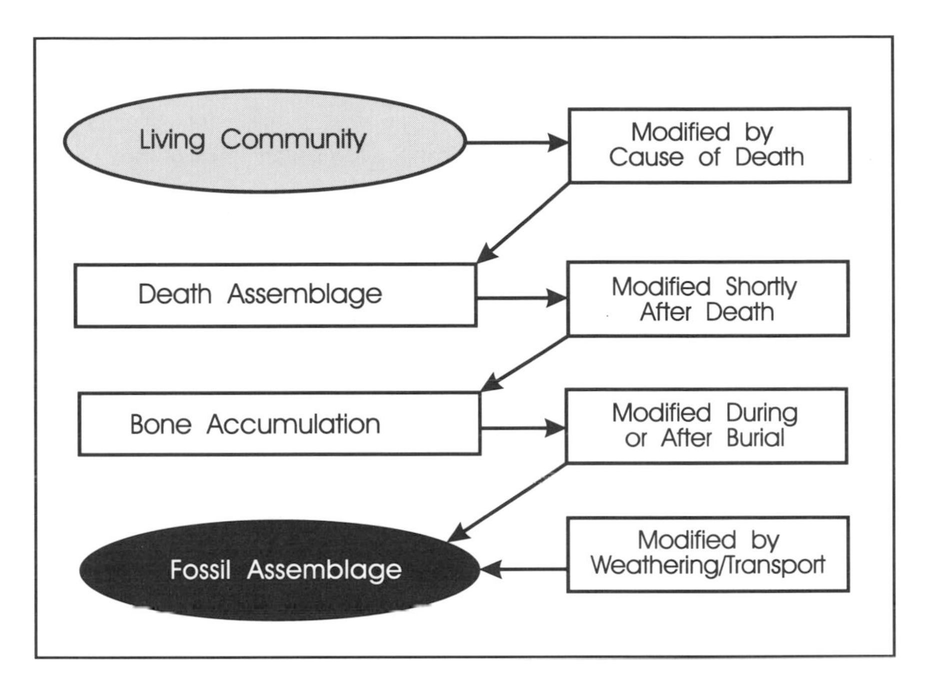

Figure 2.5. Summary of bone taphonomy processes. (Modified from Andrews, 1990.)

deposition in the lake, the wind-pollinated plants produce orders of magnitude more pollen, so they simply overwhelm the pollen rain.

Upland insects land in lakes and streams by accidentally falling in or by being blown into the water during flight. However, the proportion of upland and aquatic or shore-dwelling insects in most fossil assemblages is surprisingly well balanced.

Vertebrate skeletons (except for those of fish) are only occasionally preserved in water-lain sediments. Most good vertebrate fossil deposits have been found in caves or other natural traps. The taphonomy of vertebrate fossils is summarized in Figure 2.5. Bones may be broken or cracked when the animal dies (either by the impact of a fall into a natural trap or by predators); shortly after death, they may be affected by gnawing or breakage from scavengers or by trampling by other animals. Once a carcass is reduced to an accumulation of bones, chemical and physical weathering may come into play, and such factors as frost heaving, soil erosion, mudflows, and the like may move the bones around and mix them with other bones in a deposit.

Packrats are the chief agents of deposition in their middens. They forage for plants from a fairly narrow elevational zone centered on their nesting site. Therefore the plants incorporated into middens come mostly from the immediate surroundings of the rockshelter where the midden is preserved. The animal remains in middens (bones, scales, feathers, and insect remains) may come from a larger region, but it is probably safe to say that most of the animal species preserved as fossils in packrat middens lived within a few square miles of the midden site, if not closer.

Under some circumstances, then, fossils may lie relatively undisturbed in sediments or middens through many thousands of years and beyond. To the untrained observer, these deposits may look like worthless piles of mud or layers of black, tarry amberat. To the paleontologist, however, these repositories represent a bank vault, ready and waiting to offer up a treasure trove of fascinating clues to the history of life on this planet.

Destruction of Fossil Resources

There are a wide variety of sources for Quaternary fossils. These fossils are all around us, literally underfoot. Unfortunately, many valuable sources of Quaternary fossils are being exploited by people for other purposes. For example, peat is mined from now-shrinking deposits in North America, Europe, and Russia. Ancient fluvial deposits are mined as sources of sand and gravel or precious metals, such as gold. Although such mining activities have brought many Pleistocene fossils to light through the years, they have devoured many more. Ancient mammoth ivory has been used extensively for jewelry, buttons, and even piano keys and billiard balls. Artifacts from archaeological sites, such as arrow- and spearheads, pottery, and stone carvings, are taken away by "pot hunters." Yet these materials are the keys that unlock the book of prehistory; it is a shame when they are reduced to just another traded commodity or collectable curiosity.

Suggested Reading

Andrews, P. 1990. *Owls, Caves, and Fossils.* Chicago: University of Chicago Press. 231 pp.

Benson, L., and Thompson, R. S. 1987. The physical record of lakes in the Great Basin. In Ruddiman, W. F., and Wright, H. E., Jr. (eds.), *Geology of North America,* Volume K-3: *North America and Adjacent Oceans during the Last Deglaciation.* Boulder, Colorado: Geological Society of America, pp. 241–260.

Betancourt, J. L., Van Devender, T. R., and Martin, P. S. (eds.). 1990. *Packrat Middens: The Last 40,000 Years of Biotic Change.* Tucson: University of Arizona Press. 467 pp.

Birks, H. J. B., and Birks, H. H. 1980. *Quaternary Palaeoecology.* London: Edward Arnold. 289 pp.

Hall, S. A. 1977. Late Quaternary sedimentation and paleoecologic history of Chaco Canyon, New Mexico. *Geological Society of America Bulletin* 88:1593–1618.

Sharma, P. 1992. Packrat's liquid legacy. *Science* 255:155.

Van de Water, P. K., Leavitt, S. W., and Betancourt, J. L. 1994. Trends in stomatal density and $^{13}C/^{12}C$ ratios of *Pinus flexilis* needles during the last glacial-interglacial cycle. *Science* 264:239–242.

Wright, K. 1992. Revelations of rat scat. *Discover,* September, 64–71.

3

DATING PAST EVENTS

Once we have assembled bits and pieces of information from various types of fossils, the next step in reconstructing past environments is to fit the fossil data into a time frame or chronology. This procedure may be approached in a number of different ways, depending on what types of materials are available for dating and the interval of time we wish to study (Fig. 3.1). Accurate dating is essential to paleoecology; without it, it is impossible to determine the rates at which past environmental change took place (for example, did a climatic warming begin rapidly, or more slowly?). Accurate dating also makes it possible to determine whether past events took place at the same time across a broad region or whether those events were unrelated to each other.

In order to explain most Quaternary dating methods, we must make a brief excursion into the fields of high-energy physics and organic chemistry, the realm of **isotopes, ions,** and isomers. I have attempted to provide an overview of the major methods, explaining enough to allow the reader to grasp the idea, without delving into the details with which dating specialists must deal on a day-to-day basis. It always helps to understand the principles behind the technology on which one relies, even if that understanding is somewhat rudimentary. For additional information, please consult the suggested readings at the end of the chapter.

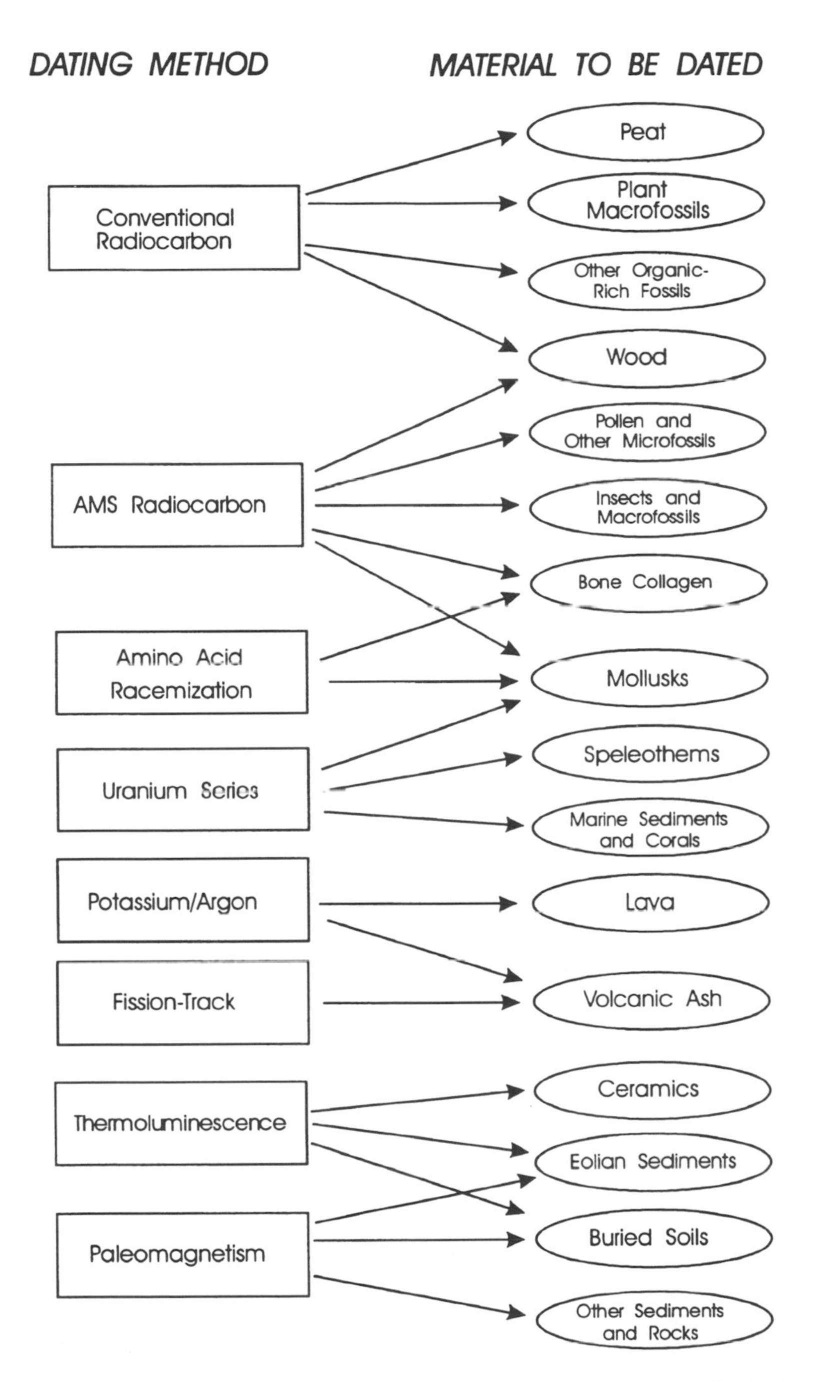

Figure 3.1. Summary of radiometric, chemical, and paleomagnetic dating methods, showing the types of fossils and other materials dated by the methods.

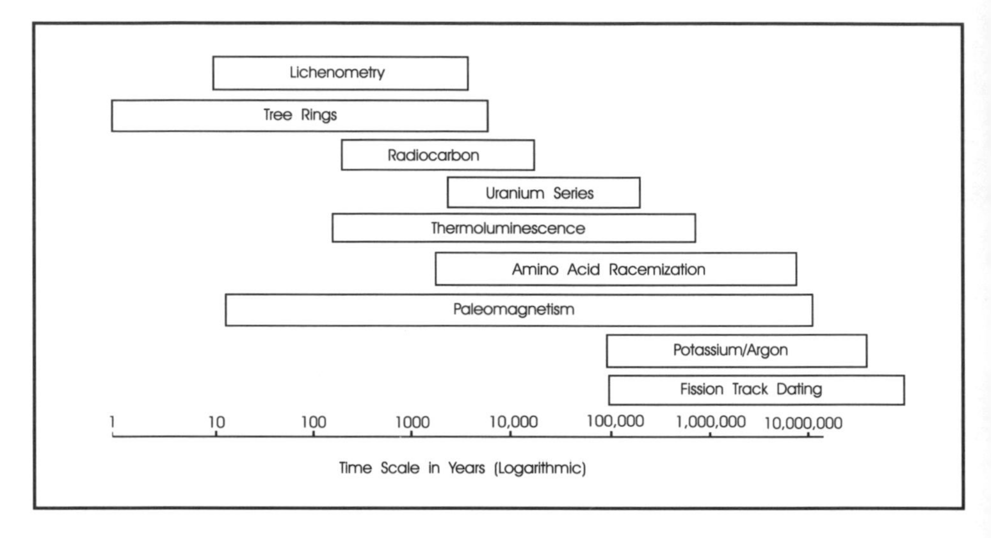

Figure 3.2. Summary of radiometric, chemical, paleomagnetic, and biological dating methods, showing the time ranges each method is capable of measuring.

Types of Dating Methods

We will look at two principal types of dating methods: radiometric and biological. Radiometric methods measure the radioactive decay of unstable isotopes of elements, including isotopes of carbon, potassium, and uranium. Thermoluminescence measures the light given off from minerals when they are heated following exposure to radiation, so it is also essentially a radiometric dating method.

Biological methods measure the growth of long-lived plants, especially trees (as expressed in tree rings) and certain species of lichens.

Each of the various dating methods is useful over a certain time span (Fig. 3.2). Some are used to date events only within the last few thousand years. Others only begin to date events 10,000–15,000 years old or older.

Radiometric Methods

The principal radiometric methods used in Quaternary studies are based on the decay of radiocarbon to stable carbon (^{14}C to ^{12}C), potassium/argon dating, and uranium series dating. Of these, radiocarbon dating is the most widely used dating method for late Quaternary fossils. The ^{14}C isotope of carbon is continuously being created in the upper atmosphere by the bombardment of nitrogen atoms by neutrons from cosmic rays from the sun (Fig. 3.3). The neutrons react with stable nitrogen atoms (^{14}N) to create a radiocarbon atom and a hydrogen atom. The

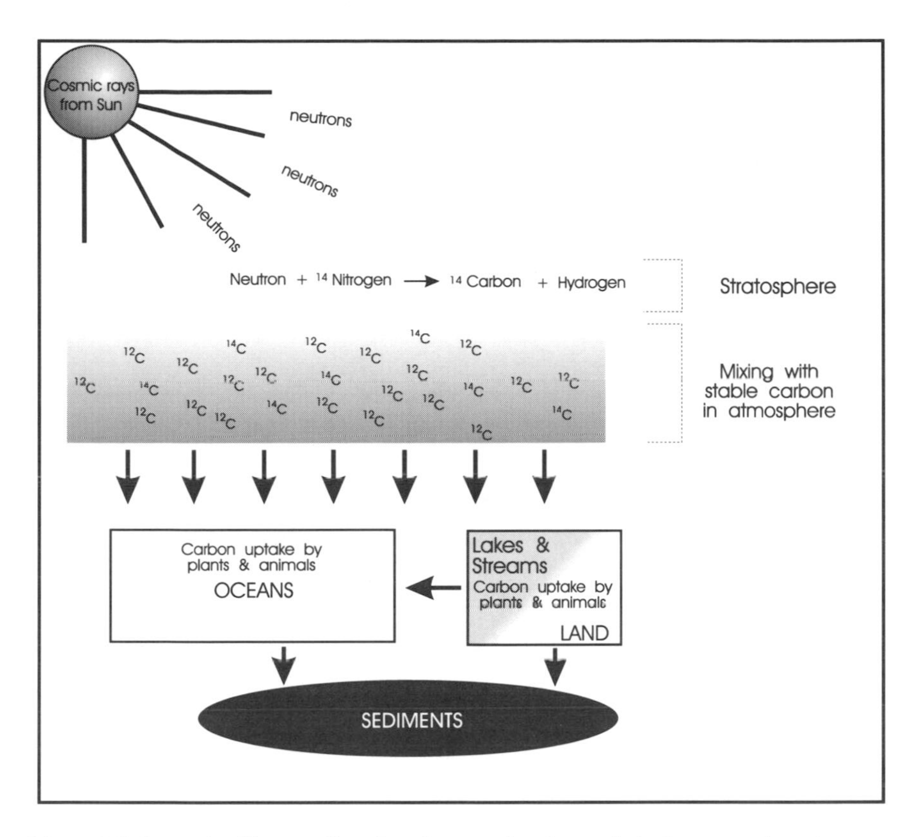

Figure 3.3. Synopsis of how radiocarbon is created and spreads in the atmosphere, its uptake by living organisms, and its final deposition in sediments. The arrows show the direction of movement of radiocarbon in the system.

radiocarbon atoms rapidly combine with oxygen to form $^{14}CO_2$, which diffuses down through the atmosphere and is taken up by plants in **photosynthesis.** The plants store the ^{14}C atoms in their tissues, along with much larger quantities of stable (^{12}C and ^{13}C) carbon. The ^{14}C atoms are continuously taken in by plants, and secondarily by the animals who eat the plants, throughout their lifetime. Even predators at the top of the food chain (such as eagles, lions, and wolves) have ^{14}C in their tissues in the same proportion to ^{12}C as is found in the atmosphere.

Once plants and animals die, they stop taking in ^{14}C, and the ^{14}C in their bodies begins to be depleted. Since the ^{14}C atom is radioactive, it begins to decay back to nitrogen (^{14}N). The rate of this decay was first determined by Willard Libby in 1955. Half of the ^{14}C atoms will have decayed to nitrogen within 5570 years. This time interval is called the *half-life* of the radioactive decay. Following the progression (Fig. 3.4), then, by 11,140 years after the death of an organism, only 25% of

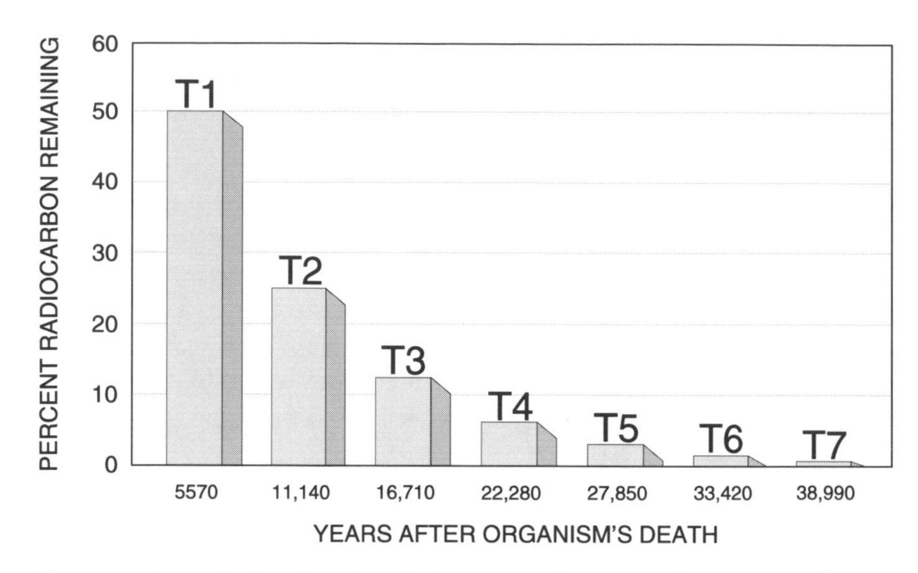

Figure 3.4. Decay of radiocarbon through time, expressed in terms of the percentage of original radiocarbon atoms remaining in a sample through seven half-lives (T = one half-life).

the original ^{14}C content will remain. By 16,710 years after death, only 12.5% of the ^{14}C will remain. By 22,280 years after death, 6.25% of the ^{14}C will be left. This exponential decrease in the amount of remaining radiocarbon can be measured to approximately 10 half-lives, or 55,700 years before present (yr B.P.). In practice, ^{14}C dating is reliable only for samples younger than 40,000 yr B.P. Beyond 40,000 years, there is so little radiocarbon left in a fossil that it becomes virtually impossible to measure accurately. Fossils older than 40,000–45,000 years yield radiocarbon ages of 40,000–50,000 yr B.P., regardless of whether they are 50,000 years old or 50,000,000 years old. Radiocarbon laboratories designate such samples as, for instance, "greater than 40,000 yr B.P."; this is then taken as the minimum age of the sample, since it could be much older.

There are now two methods of obtaining radiocarbon ages from a sample. For large samples of organic material (twigs, logs, or blocks of peat), conventional methods may be used, including either gas-counting or liquid scintillation. In the gas-counting method, the sample is burned and the gases emitted from burning (such as carbon dioxide and methane) are placed in a chamber with a detector that counts emissions of **beta (β) particles** (electrons given off in the radioactive decay process). The liquid scintillation technique involves converting the organic component of a sample into a liquid (benzene or some other organic liquid). This liquid is placed in a chamber that detects scintillations, the minute flashes of light given off when a β particle is emitted from the liquid.

Within the last decade, the development of the accelerator mass spectrometer (AMS) method has allowed much smaller samples to be radiocarbon dated. Whereas the conventional methods require samples containing at least several grams of carbon to yield reliable dates, the AMS method is able to date minuscule samples, such as individual seeds, insect parts, and tiny lumps of charcoal. For instance, the AMS method was used to obtain a ^{14}C age on single strands of fabric from the Shroud of Turin, which was claimed by some to have been the burial shroud of Christ. In fact, AMS dates from three separate laboratories showed that the Shroud of Turin was made during medieval times. Instead of measuring β particle emissions, which are an indirect measure of the amount of ^{14}C in a sample, the AMS method measures the actual abundance of the carbon isotopes (^{14}C, ^{13}C, and ^{12}C) in a sample. The atoms are accelerated in a **cyclotron** or tandem accelerator chamber to extremely high velocities. Then they are passed through a magnetic field, which separates the different isotopes, allowing them to be distinguished from each other and counted individually.

Researchers working on packrat middens in the American Southwest have used AMS radiocarbon dating extensively to obtain dates for individual species of plants found as macrofossils in packrat middens. There has been great concern about the dating of layers of packrat middens, because these deposits were not laid down at regular intervals the way lake sediments are deposited. Rather, as we have seen, the packrat accumulates plants that become incorporated into the sticky matrix of fecal pellets and urine in a rockshelter. By dating individual species of plants in midden layers, it has become possible to discern the timing of vegetation changes at a wide range of sites throughout the desert regions. In this way, researchers can gain a better knowledge of the timing of plant migrations into and out of a given region. These migrations are in turn used as indicators of regional climate change.

Thermoluminescence

Thermoluminescence (TL) is the light emitted from a mineral crystal when it is heated following the mineral's exposure to radiation. Electrons that are produced by radioactive decay (β particles) become trapped in the crystal matrix of minerals. When the mineral is heated, the electrons escape and give off light. The longer the mineral has been in the ground, collecting free electrons, the more light it will give off when heated. Since heating discharges the electrons from the crystals, it sets the TL "clock" back to zero. This makes TL useful for dating archaeological artifacts such as ceramic pottery. The clay minerals used to make a pot will discharge their TL completely when they are fired in a kiln. Any subsequent measurable TL can then be used to date the time elapsed since the ceramics were made.

Biological Methods

The biological dating methods are more "user friendly." That is, they are easier to understand and don't involve much in the way of high-powered physics or chemistry (as a biologist, I'll admit to a healthy share of bias in this statement). There are two principal biological dating methods. One is tree ring counting, or *dendrochronology;* the other (which we will not consider in detail here) is the measurement of the growth of certain lichens, or *lichenometry.*

Dendrochronology is based on the fact that, in temperate climates, trees lay down annual rings as they grow. Each ring is made up of a broad, light-colored band that represents growth in the spring and summer, and a narrow, dark-colored band that represents lessened activity in the fall. Trees tend to grow broader rings during "good" years (i.e., years with adequate warmth, moisture, and nutrients) and narrower rings during "bad" years (years of drought, disease, and cold summers). The pattern of broad and narrow rings is repeated in trees growing throughout forest regions. This allows tree rings to be matched, or correlated, from tree to tree, and even to dead trees, a process that in turn allows tree ring chronologies to be developed over greater lengths of time than the life of individual trees. Of course, tree ring researchers seek out very long-lived trees, such as bristlecone pines, from which to take cores for their studies. Fortunately, the American Southwest is home to the bristlecone pine, as well as other long-lived tree species. The cores are extracted from small holes drilled into the tree by a tool called an increment corer. These holes do not usually harm the tree; they are quickly filled in with resin, which seals the wound. Bristlecones and some other conifers may live several thousand years. By piecing together tree ring chronologies from living and dead trees (trees whose lives overlapped at some point in the past), chronologies have been extended back throughout the Holocene, or the last 10,000 years.

Tree ring dating is so reliable and accurate that it has been used as a method for calibrating radiocarbon dating. This is done by chiseling out pieces of wood from individual rings of known age and ^{14}C dating the wood. From these studies, we have learned that radiocarbon years and calendar years do not match up through the last 12,000 years. In fact, there are plateaus in radiocarbon years, especially one between 11,000 and 10,000 yr B.P. CO_2 concentrations in the atmosphere have been increasing since the end of the last glaciation, as detected in bubbles of ancient air trapped in glacial ice on the Greenland ice cap and elsewhere. There was a change in the amount of ^{14}C in the atmosphere between 11,000 and 10,000 yr B.P., causing a plateau in ^{14}C dates from this interval. This flux in ^{14}C concentrations in the atmosphere was driven by a release of CO_2 from the world's oceans towards the end of the last glaciation. The CO_2 dissolved in ocean water at that time was relatively low in ^{14}C, so it released CO_2 to the atmosphere that was likewise poor in ^{14}C . This reduced the difference between the ^{14}C age and calendar-year age of plant and animal fossils that lived in that era. Because of this phenomenon, the remains of flora, fauna, and organic sediments from the end of the last glaciation all tend to give similar radiocarbon ages. Results of tree ring studies, in combination with coral reef studies, have provided a means of calibrating radiocarbon ages back to about 20,000 yr B.P.

Paleoclimate data derived from tree rings is the best-dated evidence of changing environments in the American Southwest, because tree ring chronologies have been established in many regions. There are certain pitfalls to the use of dendroclimatology, however. Tree rings are best at indicating intervals of low rainfall, rather than those of high rainfall. When trees are stressed by lack of water, they lay down markedly narrower rings, whereas when there is an overabundance of rainfall tree ring widths may nevertheless be close to average size. During very bad years (i.e., during extreme drought), trees may cease to produce rings. This cessation leaves a gap in the tree ring record, making it difficult or impossible to determine the length of the drought. According to Jeffrey Dean, dendroclimatologist at the Laboratory of Tree-Ring Research in Tucson, Arizona, accurate dendroclimatological data are available for the southwestern region for only the last 2000 years.

Fire scars in trees are burned patches of wood produced when a tree is partially burned but survives and grows new rings over the scar. These scars can be dated to reconstruct the history of fires in a given region. Likewise, floods, landslides, and other natural disasters can damage trees and thereby leave a record of past events.

Cost Factors and Other Considerations

Many tools are used to obtain ages on sediments and fossils. Some are more precise than others, and all are applicable to only certain segments of the Quaternary time line. Often, they are used in combination. For instance, a radiocarbon age may be obtained from organic materials in silt, and a TL age obtained from the silt itself. However, none of these dating methods is cheap. The current charge for a conventional ^{14}C date is about $250, and an AMS radiocarbon date can cost more than $1,000, depending on how much work is required to prepare the sample for dating. Not many Quaternary researchers can afford to obtain more than a few dates from a given site. A common practice for organic-rich samples that are thought to be less than 45,000 years old is to obtain a radiocarbon date from a basal sample, one from near the middle of the **stratigraphic column,** and one from the top. This way the researcher can calculate rates of sedimentation that can then be used to compute the age of sediment intervals between layers that have been ^{14}C dated. When possible, researchers try to get additional dates on horizons representing times of environmental change, so that the timing of those changes can be pinned down more precisely.

Radiocarbon dating of packrat middens is necessarily more costly. Each layer of a midden must be dated to ensure accuracy, because the layers of a midden are often not laid down in an uninterrupted fashion. Rather, there may be large gaps of time between midden layers.

Suggested Reading

Berglund, B. E. (ed.). 1986. Dating methods. Chapters 14–19 in *Handbook of Holocene Palaeoecology and Palaeohydrology.* New York: John Wiley & Sons. 869 pp.

Bradley, R. S. 1985. Dating methods I and II. Chapters 3 and 4 in *Quaternary Paleoclimatology: Methods of Paleoclimatic Reconstruction.* Boston: Allen & Unwin. 472 pp.

Fritts, H. C. 1976. *Tree Rings and Climate.* New York: Academic Press. 567 pp.

Schweingruber, F. H. 1988. *Tree Rings: Basics and Applications of Dendrochronology.* Boston: Reidel. 276 pp.

4

PUTTING IT ALL TOGETHER

We have covered a lot of ground in the last three chapters. I have tried to bring the reader up to date on types of Quaternary fossils, where they are found, and how deposits are dated. In this chapter, I provide a summary of how these data are combined to reconstruct past environments. The job of synthesizing the data into a meaningful reconstruction of past environments is often the most difficult aspect of Quaternary science, but it can also be the most satisfying. After months of sieving samples, peering down a microscope, or boiling sediments in chemical baths for a pollen preparation, it is refreshing to step back from the mundane tasks and discuss the "big picture" with colleagues. The process may take weeks or months (sometimes even years) to complete, and the initial attempts to fit the data together may end in the painful decision to go out and get more or better samples before any of the important questions can be answered. The process takes patience, perseverance, and a broad outlook.

The study of ancient climates is a necessary step in unraveling ancient ecosystems, because the physical environment plays a strong role in defining ecosystems. The first level of paleoclimatic reconstruction is the process of planning the research project. This begins with a determination of which research questions are important and answerable (or at least which hypotheses can be tested). Once the research plan has been drawn up (and permission obtained from the National Park Service, if the work is to be done in a national park), the scientist can begin collecting data. Data collection involves field work to collect samples, followed by laboratory analyses of samples.

The second level of research involves converting the raw data from the laboratory into paleoclimatic data. This process ranges from simple, qualitative approaches to complex, quantitative approaches. The simplest approach is to find the modern distributions of the species in a fossil assemblage and then determine the geographic region where the modern distributions of the species in question overlap. Then the modern climate of that region can be used as an estimate of what conditions were at the time and place associated with the fossil assemblage. More sophisticated approaches involve statistical transformations of the fossil data, using mathematical formulas based on modern observations to fine-tune estimates of paleoclimates.

The third level of paleoclimate reconstruction involves combining a series of local studies into a regional synthesis. This type of study attempts to describe regional climatic patterns (such as average seasonal temperatures) over a given time interval and then proceeds to compare the patterns derived from the proxy data with theoretical climate models, such as models that reconstruct general circulation patterns in the atmosphere. Data are best synthesized by a team of researchers, each of whom approaches the research questions from a different angle or discipline (such as climatology, paleontology, or ecology). This type of interdisciplinary research (work that crosses the boundaries between scientific disciplines) is the most valuable, for it produces the most coherent, tightly focused answers, tested from many different perspectives.

The nature of interdisciplinary research is that the scientists cooperate to arrive at an answer to one question using different means, often derived from different disciplines. For instance, I recently completed a study of packrat midden insect fossils from the Chihuahuan Desert in collaboration with Tom Van Devender, a paleobotanist from the Arizona-Sonora Desert Museum in Tucson. Our aim was to compare the timing of plant and insect response to climate change at the end of the Pleistocene. Paleobotanists working in the Chihuahuan Desert had proposed a theory about the timing of this climate change, but it was based solely on one type of data: plant remains from packrat middens. When the fossil insect evidence was developed, it became clear that some species of plants (notably conifer trees) had failed to respond to the initial warming that took place about 11,500 yr B.P. The paleobotanists had been paying special attention to the timing of conifer migrations, using these species as indicators of climate change when in fact the conifer "signal" was lagging behind climate changes by 500–1000 years, as shown by the insect fossil record.

Imagination is one aspect of paleoecological research that, surprisingly, is quite essential. You may be able to assemble all the facts and figures from a fossil assemblage, but unless you can recreate a prehistoric scene in your imagination, you probably will not put the data together in a very meaningful way. This does not mean that our analyses are just a bunch of daydreams; far from it! Rather, it means that researchers have to combine their knowledge of how things work in the living world with the assembled body of fossil data in order to develop more than a superficial understanding of past events. This process involves the principle of *uniformitarianism,* formulated by British geologist Charles Lyell in 1830. This principle is summed up by the saying "the present is the key to the past." It might equally well be said that the past and present are interlocking parts of the whole, each an inseparable key to the understanding of the other. All of our modern animals and plants are just the latest generation of species that began in the distant past. If we are to understand them well enough to preserve today's populations, we need to study their history . . . their ancestral lineages that trace back hundreds of thousands of years. Thus the past becomes the key to the future.

The remainder of this chapter provides an overview of how the various bits and pieces of ancient biological and physical data are combined, or synthesized, to form major paleoenvironmental reconstructions, sometimes called "the big picture."

Reconstructing Physical Environments

The history of the physical environment is an integral part of paleoecology, because plants and animals operate in the physical world and respond as much to changes in the physical environment as they do to biotic interactions (e.g., competition for resources, predation, and parasitism). The physical environment can be broken down into three parts: the land (the geosphere), the water (the hydrosphere), and the air (the atmosphere). Obviously, the three elements interact continuously with each other. Nearly all of the energy that drives these interactions ultimately comes from the sun.

Changes in the physical environment are recorded in the features of ancient landforms. In cold regions, such features as **glacial moraines** and ancient **permafrost** features are evidence of past glacial and **periglacial environments.** In the American Southwest, these features developed only in high mountains during the last glaciation. Some lowland regions, notably in the Great Basin, developed large pluvial lakes during Pleistocene glaciations. With the exception of Great Salt Lake and some other smaller basins, these lakes dried up in the Holocene, but they left behind sediments in lake beds and ancient shorelines and downwind dune fields that are still visible along hill slopes.

Other methods of reconstructing past physical environments include studies of past lake levels and the study of the physical properties of lake sediments. Past lake levels can be deduced from ancient beach ridges or terraces that indicate the altitude of past shorelines. The quantity and type of sediments found in lake sediments can reveal information on changes in the lake itself and on changes in local environments, such as the timing of episodes of soil erosion and input of sediments from nearby glaciers.

Reconstructing Climate Change

Fluctuations in the amount of insolation (*in*coming *sol*ar radi*ation*) are the most important cause of large-scale changes in the Earth's climate during the Quaternary. In other words, variations in the intensity and timing of heat from the sun are the most likely cause of the glacial-interglacial cycles. This solar variable was neatly described by the Serbian cosmologist and mathematician Milutin Milankovitch in 1938. Three major components of the Earth's orbit about the sun contribute to changes in our climate (Fig. 4.1). First, the Earth's spin on its axis is wobbly, much like a spinning top that starts to wobble after it slows down. This wobble amounts to a variation of up to 23.5° to either side of the axis (Fig. 4.1A). The amount of tilt in the Earth's rotation affects the amount of sunlight striking the different parts of the globe. The greater the tilt, the stronger the difference in seasons (i.e., more tilt equals sharper differences between summer and winter temperatures). The full range of motion in the tilt (from left of center to right of center and back again) takes place over a period of 41,000 years.

As a result of a wobble in the Earth's spin, the position of the Earth on its elliptical path changes relative to the time of year. For instance, in Figure 4.1B, autumn and winter occur in the northern hemisphere when the Earth is relatively close to the sun, whereas summer and spring occur when the Earth is relatively far from the sun. This phenomenon is called the *precession of equinoxes.* The cycle of equinox precession takes 23,000 years to complete. In the growth of continental ice sheets, summer temperatures are probably more important than winter. Throughout the Quaternary Period, high-latitude winters have been cold enough to allow snow to accumulate. It is when the *summers* are cold (i.e., summers that occur when the sun is at its farthest point in the Earth's orbit) that the snows of previous winters do not melt completely. When this process continues for centuries, ice sheets begin to form.

Finally, the shape of the Earth's orbit also changes. At one extreme, the orbit is more circular, so that each season receives about the same amount of insolation. At the other extreme, the orbital ellipse is stretched longer, exaggerating the differ-

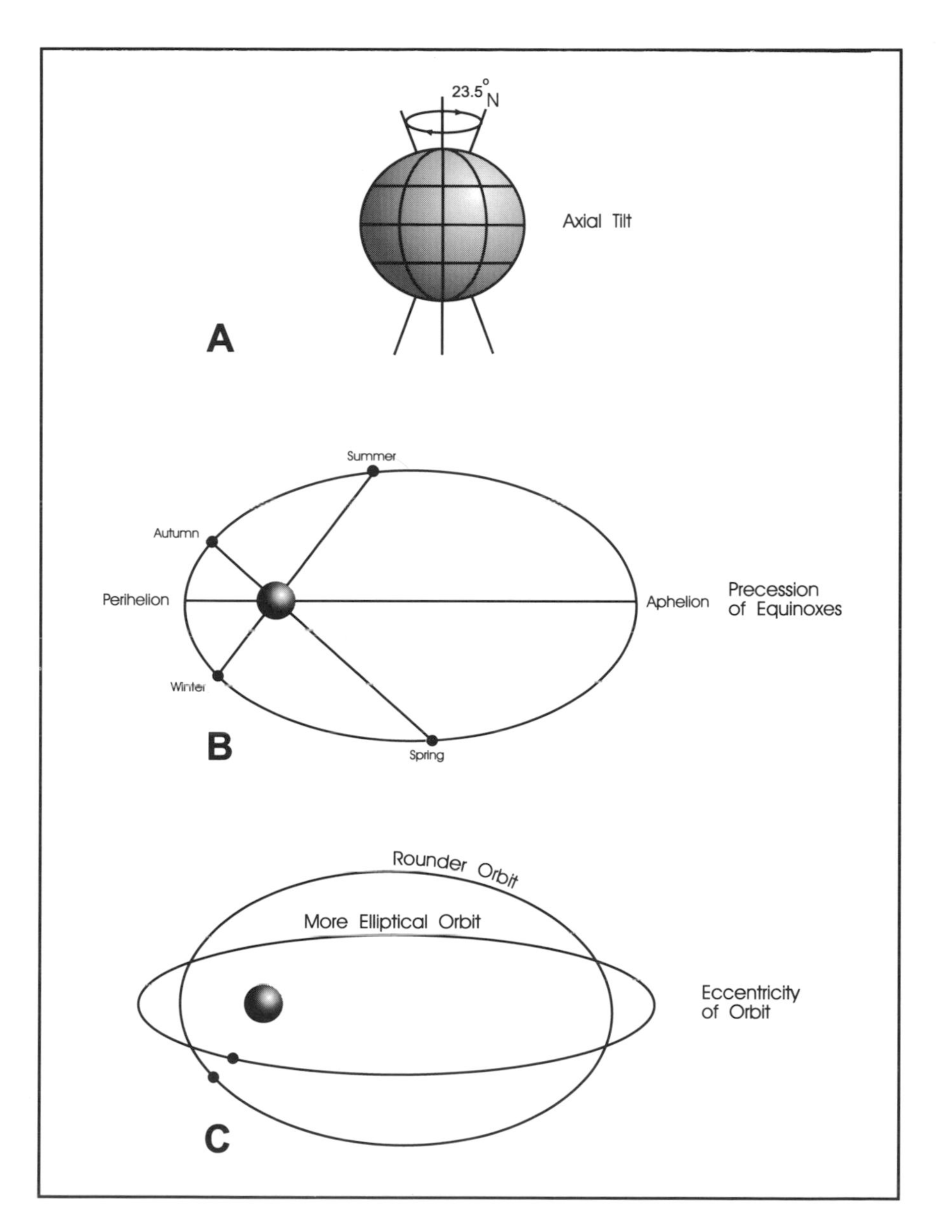

Figure 4.1. Illustration of the three major components of the Earth's orbit about the sun that contribute to changes in our climate, as described in the Milankovitch theory of ice ages.

ences between seasons (Fig. 4.1C). The eccentricity of the Earth's orbit also proceeds through a long cycle, which takes 100,000 years.

Major glacial events in the Quaternary have occurred when the phases of axial tilt, precession of equinoxes, and eccentricity of orbit are all lined up to give the northern hemisphere the least amount of summer insolation. Conversely, major **interglacial** periods have occurred when the three factors line up to give the northern hemisphere the greatest amount of summer insolation. The last major convergence of factors giving us maximum summer warmth occurred between 11,000 and 9000 years ago, at the transition between the last glaciation and the current interglacial, the Holocene.

There are other factors involved in climate change, such as volcanic eruptions and large-scale dust storms. For the most part, these phenomena trigger short-term climate change (on the scale of years to decades); however, they may interact with Milankovitch factors to draw global climate over the threshold from interglacial or **interstadial** to glacial climate.

Paleoclimatic reconstructions involve piecing together data from fossil and other physical evidence and combining it with a chronology based on stratigraphy and various types of dating. As with all of science, there are inherent problems in paleoclimatic reconstruction. One is that each type of proxy data responds to climate in its own way and at its own rate. For instance, some types of plants (especially trees) may take centuries or millennia to respond fully to a major climatic change, whereas insects may respond to the same change within a few years or decades. In this scenario, tree pollen and insect fossils from the same lake sediment samples tell very different stories. Such was the case for samples from the end of the last glaciation in Britain. The fossil insect faunas showed that summer temperatures were very close to modern levels, but the pollen records were still registering arctic tundra plants. Eventually, the conifers and then the hardwood trees arrived, but not until many centuries later.

Another problem with paleoclimatic reconstructions based on proxy data is that some data are more or less continuously deposited, whereas other data are discontinuous or spotty. For instance, a lake basin may collect sediments every year for thousands of years, whereas an adjacent glacier leaves moraines that date only to one century.

Reconstructing Ecosystems

Reconstructing ancient climates may seem difficult, but it is relatively straightforward compared to reconstructing past ecosystems. The reason is simple: plants and animals behave in much more complicated and unpredictable ways than sediments, glaciers, and frozen ground. If individual species responded independently to their environment, paleoecology would be simpler. Unfortunately, the species are constantly interacting with each other in ways that are very hard to see in a fossil record. An example of this problem is the extinction of large mammals at the end of the Pleistocene, 11,000 yr B.P. In the American Southwest, the sudden extinction of many species of large mammals took place within a few centuries of the arrival of Paleoindian hunters. Is this a coincidence, or did the hunters wipe out the large mammals? We may never know whether the animals were the victims of changing physical environments or human hunting or a combination of the two.

Like paleoclimate studies, paleoecological studies can be divided into first-level or descriptive approaches and second-level or statistical approaches. The second-level approach is gaining in popularity and success. It employs numerical and computer techniques and can sometimes reveal subtle patterns in the data that would otherwise be missed. Statistical methods can also be used to standardize the paleoecological reconstructions of several researchers working in a given region.

In order to reconstruct past ecosystems from fossil data, paleoecologists must develop studies that sample well-preserved fossils in the appropriate biological groups from the right time interval to answer the research question. Most paleoecologists are trained to identify fossils in one or a few groups of plants or animals. There are very few generalists in paleoecology, because virtually no one has the time and energy to learn to identify all the types of fossils found in a deposit. For instance, in order to identify fossil insects from sites in North America, it is necessary to develop a good working knowledge of about 100 different families of beetles, ants, and other groups! Each one of these families contains from a few dozen to a few thousand species.

Once a group of fossils has been identified from an assemblage, it is necessary to explore the ecological requirements of their modern counterparts (assuming the species in the fossil assemblage have not become extinct). Each species is adapted to a range of environmental tolerances and forms a part of a biological community. The factors that control its position within that community are described as its *niche.* The gathering of data on individual species' ecological requirements provides a good deal of paleoecological information, but a clearer picture can be obtained by reconstructing whole communities, because the community itself occupies a certain niche within a geographic region, and that niche is more narrowly defined than the niches of individual species (Fig. 4.2). For example, a given desert scrub community may be able to exist only in isolated patches on certain rocky slopes, even though several species that are a part of that community are able to live in a variety of habitats away from such slopes.

This principle is good in theory, but fossil records hardly ever preserve whole communities intact (for example, see Fig. 2.4). However, if enough elements of a

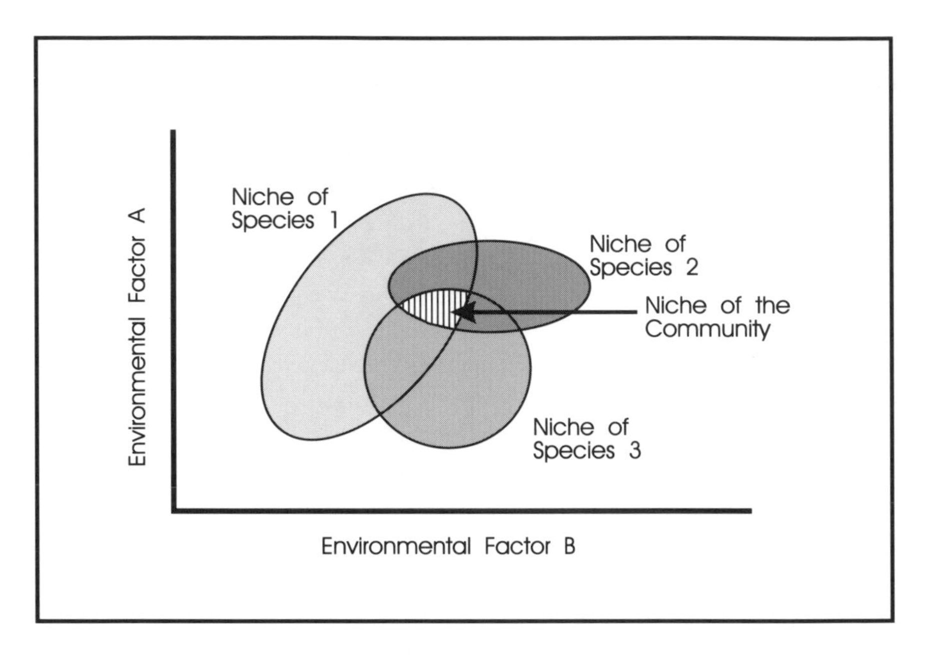

Figure 4.2. The niche of a biological community, illustrated for a hypothetical community of three species and their ecological requirements for two environmental factors. (After Birks and Birks, 1980.)

community are found in a fossil assemblage, comparisons with modern communities may be possible. This process is also strengthened by the presence of *indicator species,* that is, species that are strongly indicative of certain communities. For instance, certain bark beetles are thought to have lived only in coniferous forests because their modern descendants feed only on certain species of conifers and spend their lives under the bark of these trees. Wherever these bark beetle fossils are found, it is reasonable to assume that coniferous forests and associated boreal climate were also present, even if conifer pollen is poorly preserved or scarce in the sediments that yielded the bark beetle fossils.

Finally, the reconstruction of past ecosystems is based on the compilation of fossil assemblages from a number of regional sites representing various communities. Ancient ecosystems are described with the aid of analogous modern ecosystems. Comparing and contrasting modern analogues with ancient biotas fills in some of the gaps while at the same time showing differences in the structure and function of past and present ecosystems. No ancient ecosystem was exactly like any modern ecosystem. This is because ecosystems are constrained by the conditions of the physical environment, which has changed continuously through time. The other disparity is that past communities were made up of unique mixtures of species. Even though there can be similarities between fossil assemblages through time, species migrate, become established in new regions, and die out in others. All of this is in response to changing environments, competition between species, and changing resource availability.

Archaeology: How People Fit into the Picture

With the possible exception of causing the megafaunal extinctions, humans probably had less impact on North American ecosystems than they did on ecosystems in Europe and Asia. Certainly this was true until the Europeans arrived on the scene, beginning in 1492. European paleoecologists are often faced with the dilemma of trying to separate human modifications of past landscapes from natural changes brought about by climate change or other factors. In other words, anthropogenic (human-induced) effects on European landscapes date back several thousand years, confounding attempts to reconstruct natural environments and ecosystems. In North America, the effects of human activities are more difficult to discern, but they were probably not trivial in many regions. At the height of their prehistoric populations, there were perhaps 20–30 million Indians in North America. This many people must have affected the landscape in a number of ways, but we have yet to understand the extent and variety of their land-use practices.

Let us close this chapter by considering one of the biggest remaining mysteries in the study of North American ecosystems: the mass extinction of large mammals, or **megafauna,** at the end of the last ice age, 11,000 years ago. One school of thought holds that early hunters (Paleoindians) drove most of the North American megafauna to extinction shortly after they arrived from Asia via the **Bering Land Bridge.** Another theory suggests that the megafauna became extinct because their habitats were greatly disrupted as environments changed at the end of the ice age.

Some of the confusion that has evolved from this controversy stems from a degree of uncertainty about the timing of the arrival of humans in the New World. In the 1970s, paleontologists and archaeologists working on the Old Crow River in the northwestern Yukon Territory discovered artifacts that appeared to place humans at that site during the middle of the Wisconsin Glaciation. The artifacts from Old Crow made headlines and brought out the film crews from the National Geographic Society. Unfortunately, accelerator mass spectrometer radiocarbon dates on collagen from bones that had unquestionably been made into tools by humans showed that these artifacts were of Holocene age. Other bone "tools," such as flakes of mammoth leg bones that appear to have been fashioned into tools, in fact yielded genuine Pleistocene radiocarbon ages, but many archaeologists are skeptical that humans shaped these objects. No stone tools were found in the

Pleistocene deposits from Old Crow, just pieces of mammoth bone that may or may not have been modified by people. As of this writing, there are no unequivocal North American archaeological sites older than about 12,000 yr B.P.

Some scientists believe that Paleoindians played an insignificant role in the demise of the megafauna. They argue that the megafauna was already on its way out because of environmental factors and that the Paleoindians merely delivered the coup de grâce. Whatever caused the losses, the result was a radical reduction in the number of large mammals in the New World. This phenomenon certainly had huge impacts on all the North American ecosystems. Even though the Holocene appears to have been a time of pristine ecosystems, unaltered by human impacts in the New World, this loss of megafaunal mammals undoubtedly changed the balance of nature in ways that we have a hard time evaluating, because the ecosystems of the previous interglacial period (130,000 years ago) supported a wide range of large mammals that were absent from the Holocene. It is far more difficult to reconstruct the ecosystems of the last (Sangamon) interglacial period, because fossil sites of this great antiquity are few and far between.

The fields of archaeology and paleoecology collaborate under the banner of *geoarchaeology.* Research in this field attempts to fit archaeology more closely into a paleoenvironmental scenario. Fossil data are collected from the archaeological site or from adjacent natural deposits or from both. Moreover, geoarchaeologists use fossil data (rather than just archaeological artifacts) to help develop an understanding of prehistoric peoples and their ways.

Suggested Reading

Birks, H. J. B., and Birks, H. H. 1980. *Quaternary Palaeoecology.* London: Edward Arnold. 289 pp.

Bradley, R. S. 1985. Non-marine geological evidence. Chapter 7 in *Quaternary Paleoclimatology: Methods of Paleoclimatic Reconstruction.* Boston: Allen & Unwin. 472 pp.

Dearing, J. A., and Foster, I. D. L. 1986. Lake sediments and palaeohydrological studies. In Berglund, B. E. (ed.), *Handbook of Holocene Palaeoecology and Palaeohydrology.* New York: John Wiley & Sons, pp. 67–90.

Digerfeldt, G. 1986. Studies on past lake-level fluctuations. In Berglund, B. E. (ed.), *Handbook of Holocene Palaeoecology and Palaeohydrology.* New York: John Wiley & Sons, pp. 127–131.

Dixon, E. J. 1993. *Quest for the Origins of the First Americans.* Albuquerque: University of New Mexico Press. 154 pp.

Elias, S. A. 1995. The *Ice-Age History of Alaskan National Parks.* Washington, D.C.: Smithsonian Institution Press. 150 pp.

Elias, S. A., and Van Devender, T. R. 1992. Insect fossil evidence of late Quaternary environments in the northern Chihuahuan Desert of Texas and New Mexico: Comparisons with the paleobotanical record. *Southwestern Naturalist* 37:101–116.

Holliday, V. T. (ed.). 1992. *Soils in Archaeology: Landscape Evolution and Human Occupation.* Washington, D.C.: Smithsonian Institution Press. 254 pp.

Martin, P. S., and Klein, R. G. (eds.). 1989. *Quaternary Extinctions.* Tucson: University of Arizona Press. 892 pp.

Rapp, G., Jr., and Gifford, J. A. (eds.). 1985. *Archaeological Geology.* New Haven, Conn.: Yale University Press. 435 pp.

PART TWO

Ancient Life and Environments of the National Parks of the American Southwest

IN MOST TEMPERATE REGIONS, water-lain deposits give us the fossils necessary to reconstruct Quaternary environments. Ironically, it is the very aridity of the Southwest that has allowed us glimpses into its ancient life. The hot, dry air that permeates southwestern summer days slowly drew the moisture out of the carcasses of animals and plants. Without moisture, decomposition may be retarded or practically avoided altogether. Whereas the temperate regions of the continent have their water-lain fossil deposits, the Southwest has its mummified remains of ancient life, including hair, bones, dried tissues, and dung. This preservation has mostly taken place in the rockshelters and caves weathered out of the massive sandstone, limestone, and granite rock formations that dominate so much of the Southwest. These natural tombs have sheltered Pleistocene fossil remains, protecting them from the biological and physical forces that would otherwise long ago have scattered them to the winds. The Pleistocene fossil record of this region is therefore unique, and full of fascinating details about ancient life.

Nearly all of the Southwest was spared the effects of glacial ice during the Pleistocene glaciations. The latitude of the region was beyond reach of the continental ice sheets that formed in the arctic and spread south, and the elevation of most of the southwestern terrain was too low to foster the growth of montane glaciers. However, this region was far from unaffected by the glaciations. The climatic regime that brought ice sheets to the northern latitudes also brought cool, moist conditions to the Southwest. This change in climate did not just make the desert bloom; in many places it made the desert go away. In fact, the desert vegetation that covers most lowland regions in the American Southwest today was virtually nonexistent in the Pleistocene. In its place, different biological communities formed. The juniper-piñon woodland of the modern middle-elevation highlands spread out like a vast, green carpet across the lowlands, from the Chihuahuan Desert in the east to the last outpost of land on the west coast, Catalina Island. The mixtures of plants in regional communities were, for the most part, unlike those seen today. This statement should not surprise us greatly, because the environmental conditions that governed the mixture have likewise vanished.

It is important to remember that the vast majority of the Quaternary has been spent in glacial intervals, not in interglacial periods such as the Holocene. In fact, geologists reckon that glacial periods took up more than 90% of the last 1.7 million years. In other words, the desert ecosystems we see in the American Southwest today are the exception, not the rule. Much more typical ecosystems for this region, on the geologic time scale, are coniferous woodlands with abundant shrubs and semidesert grasses.

The animals that cavorted across the mountains, plateaus, and valleys of the Southwest during the Pleistocene were a mixture of species that still survive (although often not in the same locations as before) and those that became extinct at the end of the last glaciation. Many were giants, by modern standards. Extra-large versions of bears, wolves, lions, camels, and bison interacted with mammoths and giant ground sloths. These members of the Pleistocene megafauna and their ancestors had dominated the region for more than a million years, only to vanish at the end of the last glaciation. Evidence from dried scat of these large mammals indicates that their diet was dominated by vegetation that would be difficult, if not impossible, to harvest in large quantities today in most of the Southwest. Shrubs, herbs, and conifer needles were important elements in the diets of large herbivores such as ground sloths.

Nearly all of the megafaunal mammals (adults weighing more than 40 kg [88 lb]) became extinct in the Southwest at the end of the Pleistocene. Even species that persisted elsewhere, such as bison, died out in the Southwest. The remaining large mammal species in this region can practically be counted on one hand. They include mule deer, antelope, and mountain lions. (Coyotes, though widespread, do not fit into the megafauna category; they do not weigh 40 kg, even after consuming a large meal!)

What caused these majestic creatures to perish at the end of the Pleistocene? This question is hotly debated today, and it will likely never be settled to everyone's satisfaction. As we have seen and will again discuss in the chapters to follow, some scientists believe that human hunters wiped out the megafauna. They argue that it is no coincidence that large-mammal hunters of the Clovis culture arrived in this region at just the time when the megafauna died out. Others suggest that environmental changes alone would have been enough to bring about the demise of the megafauna. There were well-documented changes in climate and vegetation at the end of the last glaciation, but this argument does not account for the fact that the same species of megafauna had somehow managed to cope with the previous glacial-interglacial transitions of the Pleistocene. In other words, the shift from cool, moist climate and relatively lush vegetation to a hotter, drier climate and desert vegetation had taken place many times in the Pleistocene, without causing a wholesale extinction of large mammals. What made the transition to the Holocene interglacial so different?

Whatever the cause of megafaunal extinction, the few surviving species represent the decimated ranks that passed the last big environmental test. Although the Holocene fauna is greatly impoverished when compared to the Pleistocene fauna, you have to admire its members' ability to survive.

The southwestern United States has a remarkable history of biological adaptation to changing environments. The drama of prehistoric peoples did not reach its climax until the very late Holocene (i.e., the last thousand years). Because no telling of the ancient history of this region would be complete without the inclusion of these very recent events, I have stretched the time line from the other books in this series (covering the ice-age environments of Alaska and the Rocky Mountains) to include this much later chapter in our planet's unwritten history. Let us move ahead and explore the paleoecology of the Southwest, especially the regions that are now national parks and monuments. In this book, we will look at Big Bend, Canyonlands, Chaco Canyon, the Grand Canyon, and Mesa Verde.

Thanks to fossil packrat middens, the caves of the Southwest yield a unique record of changing environments as impressive as the sweep of boreal forest across the Great Lakes and the Ohio Valley during the last ice age. Not only do contemporary southwestern deserts contain peculiar plants and animals, but those plants and animals (saguaro cacti, palo verde, creosote bushes, Gila monsters, roadrunners) were not all allowed to "stay put" during the last cold stage. For instance, woodlands of juniper, piñon pine, sagebrush, and oak once overrode what is now the hot Sonoran Desert near Tucson.

Yet much of the Southwest looks today as it has for thousands of years. Life in this region is always poised on the edge of drought, so the lives of plants and animals are governed in many ways by the limited, irregular precipitation. The rocky canyons, plateaus, and mountains are little changed by human activity in many places because human populations have been so small until recently. Unfortunately, now that people are flocking to these wide-open spaces, we are belatedly discovering that these desert and semidesert regions, though they look rugged and resistant to change, are really quite vulnerable to human impact. Now, more than ever before, we need to develop a broader understanding of the ecological underpinnings of the Southwest. We must learn to savor the Southwest in order to save it. The desert biota is tough as nails; it has survived untold millennia of changing climates, but it cannot hold its own against bulldozers, asphalt spreaders, and the wheels of mountain bikes.

Although the American Southwest was explored by Spaniards, including Cabeza de Vaca and Coronado, as early as the sixteenth century, it was one of the last regions to be colonized by Anglos in their spread westward during the nineteenth century.

Early scouting parties from the eastern United States considered the whole region a desert wasteland, not suitable for farming or ranching. In one sense, their assessment was correct, because without additional water much of this region will not sustain agriculture or ranching, much less large towns and cities. Through twentieth-century technology, however, we have managed to store up much of the water that comes from the surrounding high country, then send it down drainages (natural and man-made) to water even the hottest, driest parts of the Southwest. We have also drilled deep wells to pump out ground water that accumulated during the Pleistocene (human exploitation of a nonrenewable resource is not limited to minerals and petroleum reserves). In fact, archaeologists have discovered that irrigation is not a novelty in this region. Although ancient native peoples did not construct concrete dams, they did manage to collect snowmelt and rainwater and send it to where it was most needed. Near modern-day Phoenix, Arizona, they developed an extensive system of aqueducts to send water to their fields. As many as 70,000 people have lived along the Salt River drainage in late prehistoric times, so the history of water management in the arid Southwest goes back much further than the 1920s. However, the scale of human populations, agriculture, and water use has grown exponentially since prehistoric times.

It is all too easy for people to upset the ecological balance in this arid land. When the Spanish first arrived in New Mexico and southern Colorado, they wrote of the great potential for cattle ranching, based on their observations of the great desert grasslands. In the San Luis Valley of southern Colorado, the Spanish noted grass so tall and lush that it rubbed their horses' bellies. After two centuries of ranching and agriculture, the uncultivated parts of this valley have been turned into sagebrush scrubland. The lush grasses of the past are nearly gone, clinging to marginal habitats in just a few localities. This local history has been repeated all along the Rio Grande Valley, as ranchers discovered that desert grasslands cannot sustain large herds of cattle in the same way as temperate grasslands.

In a sense, the desert grasslands documented in the nineteenth century were an artifact of 10,000 years without grazing pressure. The large grazing mammals of the Pleistocene were eliminated from southwestern landscapes at the end of the glacial era. Even bison did not remain in this region during the Holocene. Asian cattle (the domesticated cow species) were the first grazers to have a significant impact on these desert grasslands in 10,000 years.

Yet ranching is not the only activity that is having such widespread environmental impact in the Southwest. The creation of dams and huge reservoirs has changed the ecology of every affected river, as well as the quantity and quality of riparian habitats along their margins. The increase in recreational use of canyon lands in the West has also had strong environmental impacts. Desert soils are exceedingly fragile. Once compacted by the weight of human hikers, joggers, and cyclists, their delicate biological infrastructure may be damaged, and erosion will then run amok.

Fortunately, national parks preserve large tracts of southwestern wilderness nearly intact. I focus much of my discussion in this book on these parks, for it is in them that we can observe the modern ecosystems of the region, and the parks have provided many opportunities for archaeological and paleontological research into the region's past. The Antiquities Act passed by Congress during the first decade of this century gives the National Park Service the authority to protect archaeological artifacts on park lands, in hopes that the evidence of the past will not be carted off for sale to the highest bidder or end up in a private trophy case. Conservation in the national parks extends not only to modern natural resources, but also to ancient ones.

5

CANYONLANDS

Scouring and Sculpturing on the Colorado Plateau

In May 1869, John Wesley Powell led a party of ten men to explore the Colorado River and its canyons. By July, they had reached the northern end of the region now set aside as Canyonlands National Park (Fig. 5.1). As was his custom, Powell climbed to the top of the canyon to scout the lay of the land. He recorded the following impressions in his diary, dated July 19:

> And what a world of grandeur is spread before us! Below is the canyon through which the Colorado runs. We can trace its course for miles, and at points catch glimpses of the river. From the northwest comes the Green in a narrow winding gorge. From the northeast comes the Grand, through a canyon that seems bottomless from where we stand. Away to the west are lines of cliffs and ledges of rock—not such ledges as the reader may have seen where the quarryman splits his blocks, but ledges from which the gods might quarry mountains . . . not such cliffs as the reader may have seen where the swallow builds its nest, but cliffs where the soaring eagle is lost to view ere he reaches the summit. . . . Wherever we look there is a wilderness of rocks,—deep gorges where the rivers are lost below cliffs and towers and

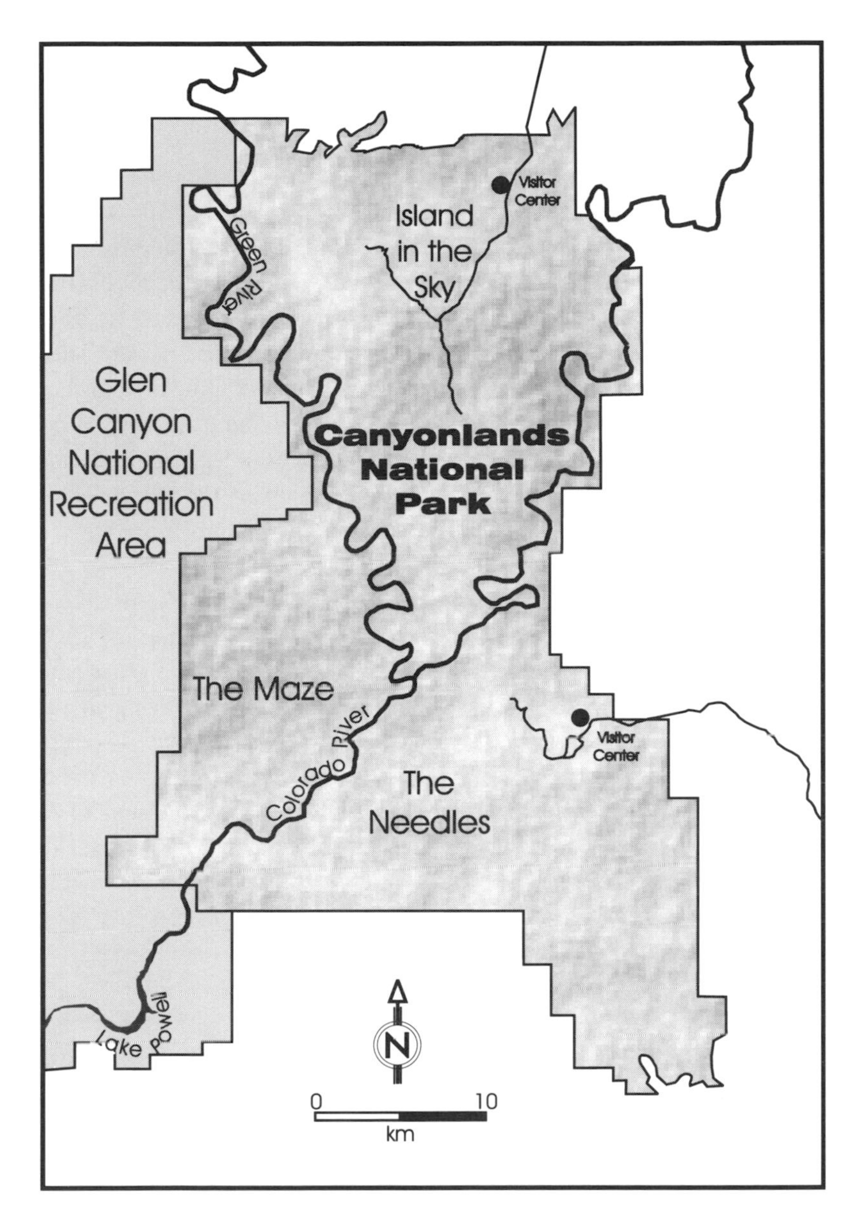

Figure 5.1. Map of Canyonlands National Park and Glen Canyon National Recreation Area.

> pinnacles, and ten thousand strangely carved forms in every direction, and beyond them mountains blending with the clouds.

Such was Powell's first impression of the "wilderness of rocks" we call Canyonlands. His voyage of discovery was plagued by mishaps and frustrations. Many have heard of the disasters that befell his party in the Grand Canyon, but his trip through Canyonlands was also no picnic. They had already lost much of their provisions thanks to repeated soakings in river water. They lost several oars when Powell's boat, the *Emma Dean,* was swamped above Cataract Canyon, and they were forced to spend many back-breaking hours portaging their boats around the numerous dangerous rapids and falls created by massive boulders that lay in the riverbed. Their lengthy stay and repeated boat-swamping in the Canyonlands region used up and spoiled vital food supplies that would be sorely missed by the time they entered the Grand Canyon.

Powell saw no mammoths on his trip down the Colorado. It may seem ridiculous to mention this, but keep in mind that when President Jefferson sent Lewis and Clark to explore the lands acquired by the Louisiana Purchase, a generation before Powell, Jefferson fully expected them to find mammoths living in the untamed lands west of the Mississippi. In fact, mammoths had been living in Canyonlands 10,000 years before Powell's arrival. Jefferson was not so far off as might be imagined, for on a geologic time scale this interval is just the blink of an eye. Let us take a closer look at this landscape and its history. As Powell discovered, it is a land that beggars description. It also happens to have had a fascinating history during and after the last ice age.

Modern Setting

The Colorado Plateau is a well-marked physiographic province covering nearly 400,000 km^2 (154,400 square miles) of the Four Corners region (Fig. 5.2). Elevations on the plateau range from 1000 to 4600 m (3300–14,000 ft). As in other parts of the Southwest, modern vegetation zones of the Colorado Plateau (Figure 5.3) do not fall into neat elevational bands across the entire region, because local bedrock types strongly affect vegetation types. The following summary represents regional averages for vegetation zones.

Canyonlands National Park lies in desert country, in the rain shadow of highlands to the south, east, and west. Mean annual precipitation in regions below 1800 m (5900 ft) is 220 mm (9 in.) or less. As already noted, the highly dissected landscape of Canyonlands does not readily lend itself to description of vegetation based on elevational zones. As in the rest of the Colorado Plateau region, local topography and soils play a large role in controlling vegetation types. Much of the

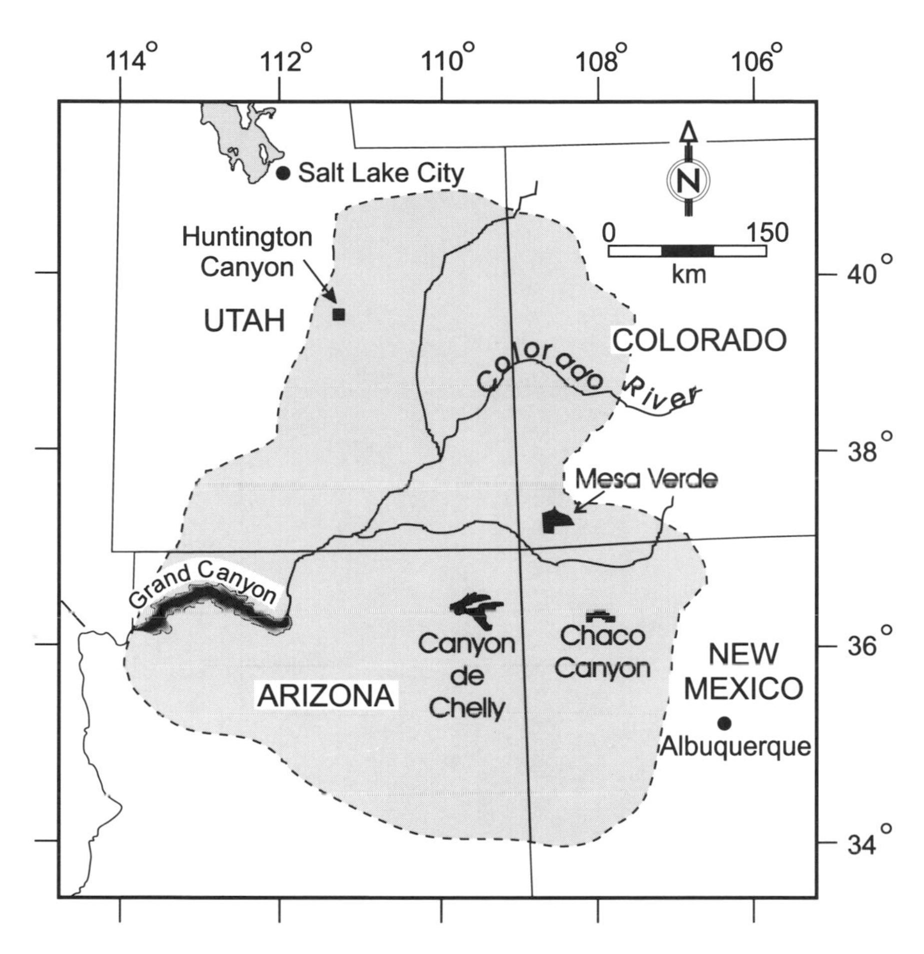

Figure 5.2. Map of the Colorado Plateau region, including portions of Colorado, New Mexico, Utah, and Arizona.

land on the plateaus overlooking canyons in Canyonlands is covered by piñon-juniper woodlands (Fig. 5.3). These woodlands are sometimes called "pygmy forests" because of their relatively low stature, compared with the more stately forests that grow at higher elevations in regional mountain ranges. The small, shrubby conifers form a transition between the arid shrublands and grasslands below and the mountain forests above.

Canyon floors are divided into riparian (near-stream) and desert scrub habitats. The riparian vegetation is more lush, with grasses, willows, and other herbs and shrubs. Along many reaches of the canyons, however, there is little or no riparian zone for the development of this vegetation, as rock walls come down to the water's edge.

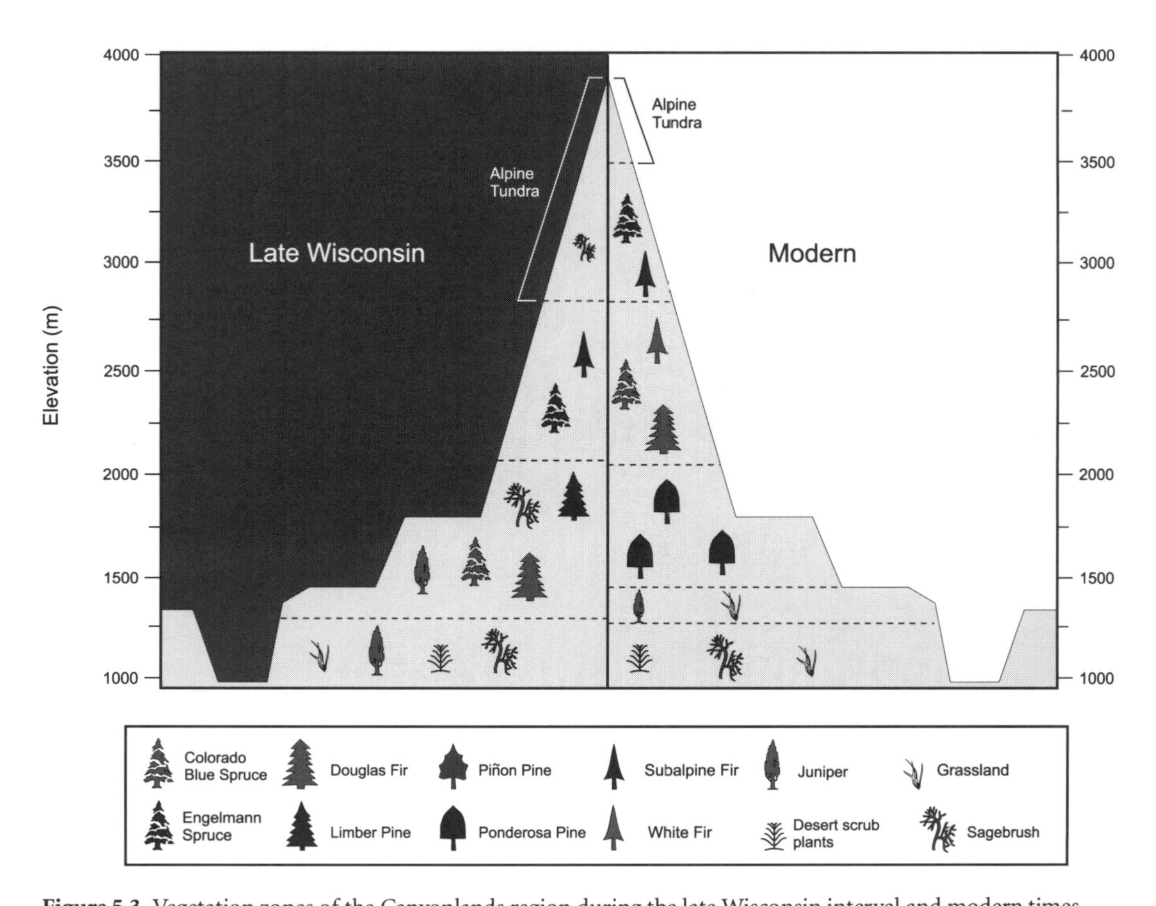

Figure 5.3. Vegetation zones of the Canyonlands region during the late Wisconsin interval and modern times. (After Betancourt, 1990.)

Wildlife in Canyonlands is perhaps less spectacular than that in other western national parks. You will not find the large herds of elk or bison that dwell in Yellowstone, nor the grizzly bears and mountain goats that roam the hills of Glacier National Park. In the rocky wilderness of Canyonlands, desert vegetation holds on to such flat places as have some soil. The sparse vegetation cannot support large numbers of animals. Nevertheless, this region is not completely devoid of animal life. Like other desert regions, Canyonlands is an area that must be taken on its own terms. You may have to get down on hands and knees to see much of the fauna here, but that does not make the wildlife uninteresting—just different from that in temperate regions.

The park is home to 65 species of mammals; one-third of these are rodents. Nineteen species of bats frequent the park. Small herds of mule deer can be found in the piñon juniper woodland, and bighorn sheep are occasionally spotted grazing on seemingly unreachable slopes. Among the rodents that live here, many are tied closely to desert habitats, including kangaroo rats, the piñon mouse, and the ubiquitous packrat.

The most conspicuous carnivore in the park is the coyote, the resourceful predator that is viewed by the native peoples of the southwest as a wise prankster. Regional foxes include the gray fox and the kit fox, a true desert fox that preys on jackrabbits, mice, kangaroo rats, cottontails, and birds. Another regional predator is the ringtail; this species ranges from Mexico up to northern Utah. A secretive, nocturnal animal, the ringtail preys on rodents, lizards, birds, and insects. The largest carnivore frequently seen in the park is the mountain lion, a solitary, wide-ranging cat whose home range may cover thousands of square kilometers.

Lizards and snakes are also important components of the regional biota. Rattlesnakes keep hikers on their toes during the summer months. According to herpetologists Jim Mead and Christopher Bell, there are 61 species of reptiles and amphibians living on the Colorado Plateau and Great Basin. Only a few of the reptiles that live on the Colorado Plateau are poisonous, including the sidewinder, speckled, and western rattlesnakes and the Gila monster. The Gila monster is mostly found in the Sonoran and Mojave deserts, but it ranges into the lowlands along the Colorado River at the western end of the Grand Canyon.

The modern fauna is but a poor remainder of what lived here during the last ice age, however. Cooler, moister conditions fostered the spread of more bountiful vegetation, which in turn allowed a far greater variety of large animals to survive here until about 10,000 years ago. We will start our prehistoric tour of the park with a look at that ancient vegetation and the climates that spawned it.

Glacial History of the Colorado Plateau

The last glaciation of the Pleistocene is called the Pinedale Glaciation in the Rocky Mountain region. The Pinedale Glaciation began after the last (Sangamon) interglacial, perhaps 110,000 yr B.P., and included at least two major ice advances and retreats. These glacial events took different forms in different regions. Real glaciation took place only on the edges of the Colorado Plateau, in adjacent mountain ranges. During late Pinedale times, glaciers that filled high valleys in the San Juan Mountains of southern Colorado flowed down to fill adjacent valleys, including the Animas River Valley that heads near Silverton. According to U.S. Geological Survey geologist Paul Carrara, the Animas Glacier flowed down to what is now the townsite of Durango, where its terminal moraine can still be seen. Smaller glaciers flowed down valleys in the La Plata Mountains, not far from Mesa Verde. These mountain glaciers were up to 16 km (10 miles) long, but they did not spill out onto the adjacent lowlands, such as the Mancos Valley. Most of the Colorado Plateau region did not have an ice cap during the Pinedale Glaciation.

This glaciological history in no way implies that it wasn't cold on the unglaciated parts of the Colorado Plateau during the last glaciation. Exciting new evidence has recently become available for Pinedale environments and biota of this region, based on scattered but significant fossil finds. For instance, researchers digging near an archaeological site by the Dolores River in southwestern Colorado unearthed a partial skull, vertebrae, and ribs of an extinct Pleistocene musk-ox, *Ovibos* (the musk-ox remains were far older than the archaeological artifacts at the site). Collagen extracted from some of these bones has been radiocarbon dated at about 16,000 yr B.P., right in the middle of the last glacial episode in the nearby Rockies. Today, musk-oxen are found almost exclusively in arctic tundra regions of Greenland, Canada, Alaska, and Siberia. Although the Pleistocene musk-ox has become extinct, it seems likely that it, too, was adapted for life in arcticlike climates with tundra vegetation.

The remains of another cold-adapted animal, the pika, have been found at low-elevation sites near the La Plata Mountains in the Four Corners region. Pikas live today only at, or just below, the alpine tundra region. So the vertebrate fossil record indicates that during the last glaciation the Four Corners region was probably substantially cooler than it is today.

The late Pinedale glaciers advanced to their outermost extent between about 20,000 and 18,000 yr B.P. but were still quite substantial at 16,000 yr B.P. In the southern and central Rocky Mountains, fossil evidence from insects and pollen suggests that even lowland regions experienced climatic conditions that were substantially colder than those at present. For instance, average summer temperatures were probably as much as 10°C (18°F) cooler than they are today. Because of the Milankovitch cycles, summer insolation reached a minimum level at 18,000 yr B.P., then began slowly to increase.

Farther west, in the Great Basin region, immense lakes filled many closed basins. Throughout the western United States, herds of mammoths, llamas, camels, and Pleistocene horses roamed the lowlands and plateaus. All of these creatures became extinct by the end of the Pleistocene. According to geologists Larry Agenbroad and James Mead of Northern Arizona University, mammoths became extinct on the Colorado Plateau sometime shortly after 11,300 yr B.P. No direct associations have yet been found between mammoths and Paleoindian hunters in this region. In other words, no mammoth kill sites similar to those seen in southeastern Arizona and eastern New Mexico and Colorado have been documented, and no human artifacts have been found associated with mammoth remains.

Regional climates of the Colorado Plateau are today strongly affected by topography, and such was also the case in the late Quaternary. The interior regions fall within the rain shadow of highlands to the south, east, and west. In general, both Pleistocene and modern vegetation of the plateau suggests wetter conditions than the vegetation of similar elevations farther west. Large-scale changes in this region's climate are thought to have been brought about by shifts in the dominant patterns of the jet stream, the main circulation pattern in the upper atmosphere for the northern hemisphere.

By the end of the Pinedale Glaciation, about 11,000 yr B.P., regional climates were changing rapidly. Summer temperatures rose to near the modern level, and the landscapes of the Colorado Plateau became more arid as the atmospheric circulation patterns that brought increased precipitation during Pinedale times began to dissipate. In the San Juan Mountains of southern Colorado, fossil insect data suggest that summer temperatures reached modern levels before 10,000 yr B.P. and became warmer than modern temperatures by about 9600 yr B.P. The modern vegetation zones seen on the Colorado Plateau began to take shape, but this was an ongoing process that required much of the Holocene to complete.

Late Pleistocene Vegetation History

The following reconstruction of ancient vegetation and environments has been brought to you by packrats, those lively little plant collectors who, fortunately for paleontologists, have an unswerving dedication to the credo "never throw anything away." Rockshelters abound in Canyonlands, where weathering gouges depressions, overhangs, and clefts in the sandstone. Many rockshelters were apparently occupied more or less continuously by a series of packrat residents throughout the late Pinedale, providing us with the dried remains of local vegetation,

Figure 5.4. A packrat midden in a rockshelter, Arches National Park, Utah. (Photograph by Saxon Sharpe.)

trapped in amberat (Fig. 5.4). Yet reconstructions of ancient environments in this region are not limited to those regions where middens have been found. There are not as many permanent lakes, ponds, or bogs on the plateau as are found in other parts of North America, but there are sufficient long-lived bodies of water to allow detailed paleoenvironmental studies to be carried out, especially with pollen. Researchers have therefore relied on the packrat midden record to extend paleoenvironmental reconstructions to new regions where lakes are scarce and to fill in many vital details of paleoenvironmental history. More than 180 packrat middens have been analyzed from sites on the Colorado Plateau.

During the last glaciation, the Canyonlands region received substantially greater amounts of moisture as the jet stream shifted course, bringing Pacific moisture that today falls on higher latitudes. Insolation was markedly lower at 18,000 yr B.P. than it is today, so the whole of North America experienced cooler climate at that time. Paleoecologists have concluded that during late glacial times upper **treeline** on the plateau was significantly lower than it is today, and alpine tundra regions may have had more sagebrush cover than is found there today. Alpine tundra is found today in only three high mountain regions of the plateau, at elevations above 3660 m (12,000 ft). Late glacial coniferous forests (dense stands of trees) extended down to 2000 m (6550 ft) elevation in **mesic** canyon habitats. Coniferous woodlands (conifer trees growing in an open, parklike landscape) extended down to 1300 m (4250 ft) elevation on rocky sites near the San Juan Mountains and in Glen Canyon. Canyons and other lowlands below that elevation were covered by desert scrub with junipers and sagebrush.

The plateau country dominated by piñon-juniper woodland today supported a more diverse woodland during the last glaciation, including limber pine, juniper, Colorado blue spruce, Douglas-fir, and sagebrush. The most common association was Rocky Mountain juniper, limber pine, and Douglas-fir. This woodland vegetation covered all but the lowest elevations in the park at that time. Curiously, piñon pine remains have not been found in Pleistocene middens in the Canyonlands region or elsewhere in southeastern Utah. Paleobotanist Julio Betancourt has suggested that this absence may be due to the fact that the tree's upper elevational limit during the glaciation fell beneath the lowest elevations of this region. It grew at elevations below 1700 m (5600 ft) in southern New Mexico during the last glaciation and did not migrate onto sites in the Four Corners region (such as Chaco Canyon, Canyonlands, and Mesa Verde) until after 10,000 yr B.P. The cooler conditions experienced at the higher latitude of Canyonlands may have kept piñon pine out of this region. Ponderosa pine was also apparently absent from this region during the last glaciation, even in habitats where it is now common. Limber pine seems to have taken its place in some regions. This is a reversal of roles. Limber pine was abundant in glacial times and is relatively rare today; ponderosa pine was rare in glacial times and is abundant today. Such changes of fortune are common occurrences over geologic time. Today's success story was yesterday's marginal species, and vice versa. On the whole, species that thrive during glacial periods have enjoyed longer intervals of success than species that thrive during interglacial periods. As we have seen, glacial intervals comprise roughly 90% of the last 1.7 million years, so these have been the rule while warm interglacial periods have been the exception during the Quaternary Period.

Vegetation was not static throughout the last glaciation. Full-glacial environments (22,000–18,000 yr B.P.) were colder than those that preceded them as well as those that came after them. These changes have been documented in several series of radiocarbon-dated packrat midden fossil assemblages. For instance, a rockshelter at 1820 m (5900 ft) elevation in Natural Bridges National Monument yielded plant remains that range in age from greater than 39,800 to 23,400 yr B.P. The plant macrofossils preserved in older layers of the midden include abundant remains of spruce, Douglas-fir, limber pine, and wild rose (*Rosa woodsii*). Younger layers include these same plants, plus dwarf juniper (*Juniperus communis*) and Rocky Mountain juniper (*Juniperus scopulorum*).

Late glacial midden assemblages, dating from 14,000 to 10,000 yr B.P., show a series of vegetational changes in the Canyonlands region. Colorado blue spruce became more abundant at some sites during this interval, at the expense of limber pine. At a rockshelter at Natural Bridges National Monument, a midden sample dated 9700 yr B.P. contained abundant Douglas-fir and traces of spruce and limber pine. This finding suggests that the regional climate was still cooler and moister than it is today, although fossil insect evidence from elsewhere in the Southwest indicates that woodland species may lag behind changes in climate for many centuries, especially at the end of the last glaciation.

A series of rockshelters from the Escalante River basin in southeastern Utah yielded packrat middens that were used by paleobotanist Kim Withers and geologist James Mead to reconstruct past vegetation. Midden samples older than 11,000 yr B.P. contained species that live at higher elevations today, such as Douglas-fir, spruce, and mountain mahogany, as well as plants that need quite a bit more moisture than is available at the sites today, including wild rose and water birch. After 11,000 yr B.P., both the cool-adapted and the moisture-loving plant species disappear from the midden records of this region. The composition of the vegetation continued to change in the early Holocene; by 7000 yr B.P., the only species represented in the middens were Gambel oak and prickly pear cactus. The authors attribute the increased available moisture in late glacial times to higher stream levels and increased amounts of groundwater, both of which tapered off in the early Holocene, although the soils remained moister than they are today. The alcoves that housed the middens supported a mixture of vegetation until stream levels dropped off to the point where the alcoves became too dry to support plant life.

During the Holocene, the lower elevational limit of conifer woodlands shifted upslope. In many places, this vegetational shift was not completed until about 6000 yr B.P., even though regional climates undoubtedly began warming and drying much earlier. At the transition between the last glaciation and the Holocene, nearly every species of plant was either coming or going, shifting uphill or down, in an enormous reshuffling of biological communities. Some species migrated slowly, taking hundreds or even thousands of years to become established where they are today. Others moved rapidly, shifting their ranges in a few decades or centuries. In some cases, species were not removed from a given region until a better-fit competitor arrived. The current biological communities, though they may appear to be stable, balanced associations in tune with environmental conditions, are just the latest reshuffling of biological players on the ecological stage. Communities only appear stable to us because our lifetimes are too short to take note of changes on the ecologically important time scales, which are centuries to millennia. This is one of the most important lessons to be learned from paleoecology: all biological communities are ephemeral. Associations of plants and animals are in a constant state of flux, if viewed on a meaningful time scale.

Late Pleistocene Mammals of the Colorado Plateau

The megafaunal mammals that once roamed the Colorado Plateau were a fascinating group of creatures, well adapted to ice-age environments in this region. Because of the abundance of dry caves and rockshelters on the plateau, the vertebrate fossil record for the late Pleistocene is remarkably well preserved, giving us much information about these animals. If it weren't for this extensive record, we probably would never have guessed that this now-arid region could have been home to such a wide variety of large mammals during the Pleistocene. The dry valleys that today support only a rather meager-looking desert scrub vegetation do not look like good habitat for mammoths, horses, sloths, musk-oxen, and camels. But in the Pleistocene everything was different. Grasses were more common; shrubs were more luxuriant; the whole scene may have looked more like the modern African savannah. After all, you do not have to tip the moisture scale all that far from arid to moist to change desert scrub into semidesert grassland or steppe. Even though the Pleistocene landscapes of the Southwest were not exact matches for these modern ecosystem types, they were not far off. The fossil evidence from large grazing and browsing mammals requires us to consider that highly productive plant communities existed in the Southwest; otherwise the megafaunal mammals would not have lived there for tens of thousands of years.

Two members of the elephant group (proboscideans) lived on the plateau during the last glaciation. These were the Columbian mammoth and the American mastodon. The Columbian mammoth was roughly the size of the modern Asian elephant. It had long tusks that curved back on themselves, forming a shape like the letter J (Fig. 5.5). The Columbian mammoth was not as hairy as its northern relative, the woolly mammoth, but the latter species was adapted to very cold climates, whereas the Columbian mammoth lived in more temperate regions. Mammoth teeth consisted of a set of large, flat, grinding molars. These teeth had row after row of tightly spaced grinding surfaces, meant for chewing grasses and other herbs. Like the modern elephants, mammoths probably consumed huge amounts of vegetation every day, a diet that, once again, argues for a productive ecosystem with abundant vegetation of grasses, herbs, and shrubs.

We have learned quite a lot about the mammoth's diet on the Colorado Plateau from an almost unbelievable source. If you were to read a headline in a tabloid that proclaimed "Huge Quantities of 15,000-Year-Old Mammoth Dung Found in Utah Cave" or "Scientists Discover Ancient Elephant's Bathroom," you would probably place these headlines on a par with "Elvis Found Living on Mars" or "Woman Gives Birth to 20-lb Grapefruit." Yet, in spite of its improbability, this "mammoth bathroom" discovery is exactly what took place a few years ago. A large cave in Glen Canyon National Monument, called Bechan (large excrement) Cave by the

Figure 5.5. Megafaunal mammals of the Colorado Plateau region during the last glaciation. 1, Columbian mammoth; 2, Shasta ground sloth; 3; Yesterday's camel; 4, North American lion; 5, long-horned bison; 6, Harlan's ground sloth; 7, Pleistocene horse; 8, pronghorn antelope; 9, saber-tooth cat; 10, Harrington's mountain goat; 11, gray wolf. All of these except the wolf and the pronghorn antelope became extinct at the end of the Pleistocene.

Navajos, preserved an incredible bed of dried Pleistocene dung, estimated at 300 m^3 in volume (Fig. 5.6). The floor of the cave was literally covered in dung, including that of mammoths (Fig. 5.7), ground sloths, and at least six other species. Geologists Jim Mead and Larry Agenbroad and their colleagues have described the large, round mammoth dung balls, called boluses. These boluses were almost exactly the same size and shape as those of modern elephants. Did a herd of mammoths use the cave as their bathroom? We will never know why so much dung was deposited there, but it is a paleontological treasure trove, one that could only be preserved in the type of arid climates found in the American Southwest. Perhaps mammoths and ground sloths deposited large quantities of dung in caves throughout North America. If so, we would not know about it, because in nearly all other regions of the continent, the dung would have long since decomposed.

The dung boluses from Bechan Cave contained well-preserved plant macrofossils, which give us vital information about the diet of these animals. Once again the fossil record takes us from the realm of speculation into the world of cold, hard facts. We can say with certainty that mammoths on the Colorado Plateau ate large quantities of grass, because the mammoth dung in Bechan Cave was dominated by grass remains (making up as much as 95% of its composition). The remains of woody plants made up the rest of the dung macrofossils. These included twigs of saltbush, sagebrush, birch, and blue spruce. So, although mammoths in the American Southwest enjoyed a little variety in their diet, like their modern counterparts in the Old World they subsisted mostly on grasses.

Finds of fossil dung may seem uninteresting, or even disgusting, to the uninitiated. To gain an appreciation for their worth to science, it is necessary to see these fossils as the paleontologist sees them. When a medical doctor wants to evaluate your general health and condition, you are asked to provide stool and urine samples. These excreta provide valuable information about the biology of the human medical patient; likewise, fossil excreta provide vital clues about the lives of ancient animals. An anthropologist working on the excavation of a medieval site in the city of York, England, discovered human feces associated with the artifacts. The human **coprolites** happened to be found in an excavation under a bank building, and the anthropologist caused a stir in the local media when he declared that, to him, these remnants of past human life were more valuable than the contents of the bank vault.

Figure 5.6. Bechan Cave, Utah. (Photograph by Emilee Mead, in Nelson, 1990.)

Figure 5.7. Mammoth dung boluses from Bechan Cave. (From Mead et al., 1986.)

Figure 5.8. The dung beetle, *Aphodius fossor,* found in mammoth dung from Bechan Cave.

One of the fondest wishes of any vertebrate paleontologist studying a particular species of extinct animal would be to travel back in time to observe that animal's behavior, especially its feeding habits. The food that an animal eats is one of the main components that fit the species into its niche in a biological community. We have access to fossil dung of just a handful of the thousands of species that have roamed this planet in ancient times. A remarkable percentage of that fossil dung comes from Pleistocene mammals that once lived on the Colorado Plateau.

It so happens that the mammoth dung bed had much more to tell us. Not too surprisingly, the dung contained the remains of dung beetles. I was keen to examine these specimens, because I wanted to test a theory. Modern African and Asian elephants (the nearest living relatives to the Columbian mammoth) have their own cadre of dung beetles that feed either predominantly or exclusively on their dung. Did the mammoths of North America have their own dung beetles following them around? If so, what has become of them? If there were North American mammoth dung feeders living in the Pleistocene, I reasoned, they either became extinct when their source of dung disappeared (i.e., when the mammoths died out) or had to adapt quickly to new varieties of dung. The results of my little dung beetle study yielded a third solution, one that I had not thought of. The dung beetle species that I found in the Bechan Cave mammoth dung turned out to be one that is still alive and well and living in Eurasia. The fossil specimens match an Old World species, *Aphodius fossor* (Fig. 5.8). It feeds on a variety of animal dung today; it is considered a dung "generalist." At the end of the Pleistocene, it apparently died out in North America, but Eurasian populations survived, eating all kinds of herbivore dung.

The American mastodon belonged to a more ancient line of proboscideans. It was slightly smaller in stature than the Columbian mammoth and had shorter, straighter tusks. The most notable feature of mastodons was their cone-shaped

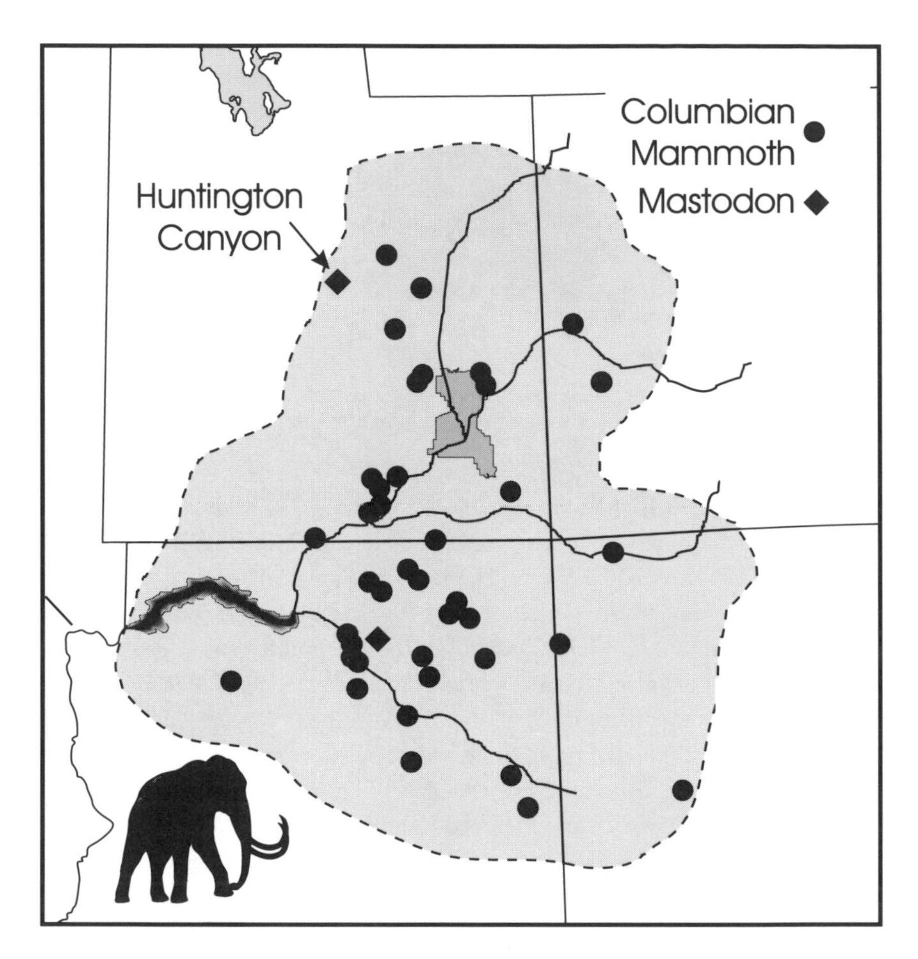

Figure 5.9. Map of mammoth and mastodon fossil sites on the Colorado Plateau. (After Nelson, 1990.)

teeth, used for chewing conifer twigs and other woody vegetation. These animals were found mostly in coniferous forests and woodlands, where they browsed on the trees and shrubs. Although mammoth remains have been found at many localities on the Colorado Plateau, mastodon fossils have only been found at two places: Huntington Canyon in Utah and the Little Colorado River in Arizona (Fig. 5.9). Mastodon fossil localities are much more common in other regions, where conifer forests were more abundant and luxurious. Again, if fossil remains of mastodons had not been found on the Colorado Plateau, their existence there in the late Pleistocene would probably never have been guessed by researchers working on proboscideans. Although mastodons may not have been as abundant here as elsewhere, their very presence reinforces the paleobotanical evidence for the importance of conifers in southwestern landscapes. If conifers had been rare or absent from lowland regions there, mastodons simply would not have lived there.

The American mastodon became extinct about 11,000 yr B.P. The story of the passing of the Columbian mammoth from the Colorado Plateau is a little more complicated. One episode of that story also took place at Huntington Canyon, near the edge of the plateau in central Utah (Fig. 5.2).

In 1988 an earthen dam was in need of repair at Huntington Canyon. The base of the dam had been built on an unstable surface: clays laid down in a late Pleistocene pond. As construction workers bulldozed the earth away from the base of the dam, they uncovered a bone bed in the clay. The bed contained the remains of a Columbian mammoth. This discovery was remarkable for several reasons. First, the site is 2730 meters (8950 ft) above sea level. This makes it perhaps the highest elevation recorded for mammoth in North America. Second, the mammoth skeleton was preserved virtually intact, with all the bones lying where the animal had died (Fig. 5.10). Based on the way the bones were positioned in the mud, paleontologists Dave Madsen and Dale Gillette put together the following scenario of its last day. The mammoth was old and weak. It found water in a small pond, ringed by dense vegetation, and waded into the water to drink or perhaps eat some tender aquatic plants. But when the old mammoth got out beyond the

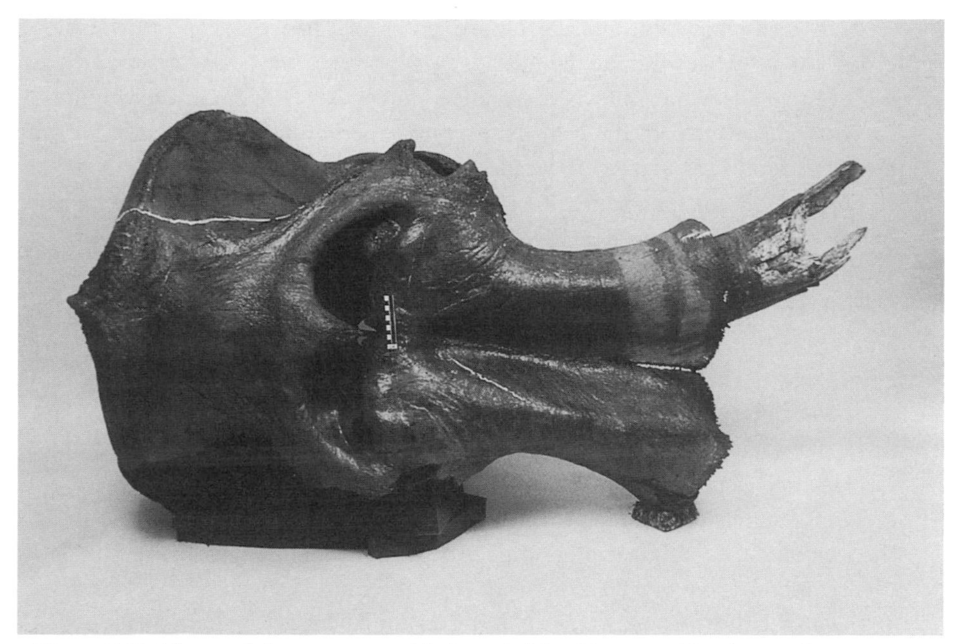

Figure 5.10. Mammoth bones preserved in late glacial sediments at Huntington Canyon, Utah. (Photograph by David Madsen, courtesy Utah Geological Survey.)

water's edge, it got hopelessly stuck in the mud. Perhaps if it had been younger and stronger, it might have freed itself. But it was old and feeble and its efforts to pull its legs out of the mud only mired it more deeply. Finally, exhausted, it slumped down in the muddy water and died.

For reasons that we will never learn, no scavengers larger than crows got to the mammoth carcass in the pond. If larger carrion feeders (wolves, coyotes, or foxes) had come to the carcass, they would surely have dismembered it and scattered the bones over a wide area. But this did not happen. Perhaps the larger scavengers were afraid of getting stuck in the mud themselves. Perhaps the water where the mammoth died was too deep for them. Whatever the reason, the mammoth carcass was not subjected to this type of scavenging. There were no gnawing marks on the bones, and the carcass stayed in place. The bones of a short-faced bear were also found at this site, but the bear apparently had no part in bringing down the mammoth or in feeding on its carcass.

Sometime after the mammoth died, the top of its head decomposed, leaving a cavity where the brain had been. That cavity filled with pond mud. Thanks to that improbable set of circumstances (mammoth dies in pond; scavengers leave it alone; braincase fills with sediment; sediment gets buried by clay for more than 10,000 years until discovered by people interested in paleontology), I was able to examine the mud that was scraped out of the mammoth's skull. Opportunities to study such remarkable finds as this are few and far between. For a paleontologist, it is a little like winning the lottery. In both cases, one must beat incredible odds to gain a treasure. But whereas a lottery prize may be spent in a few months or years, discoveries such as the Huntington mammoth will probably go on yielding scientific dividends for decades to come.

Surprisingly, the mud from the mammoth skull contained quite a wide variety of insect fossils (this was a relatively small sample, compared to the usual bulk samples with which I deal). The mixture of beetle species in this unusual assemblage indicated climatic conditions very close to the modern ones for this elevation and region. Some of the insect fossils were dated by the accelerator mass spectrometer (AMS) radiocarbon method, yielding an age of 9900 yr B.P. They turned out to be younger than the mammoth itself, whose bone collagen yielded an AMS date of 11,300 yr B.P. The mammoth died a few centuries before the last of its kind. This is a sobering example of the importance of direct radiocarbon measurement on organic material used for paleoclimatic interpretations. Without the ^{14}C dates there would be no strong stratigraphic evidence for thinking that the beetles and the mammoth were not the same age. How did mud that was several centuries younger than the mammoth end up filling its skull? That is a mystery that remains to be solved. Then again, all good scientific investigations tend to answer some questions while generating a whole new batch of questions in the process.

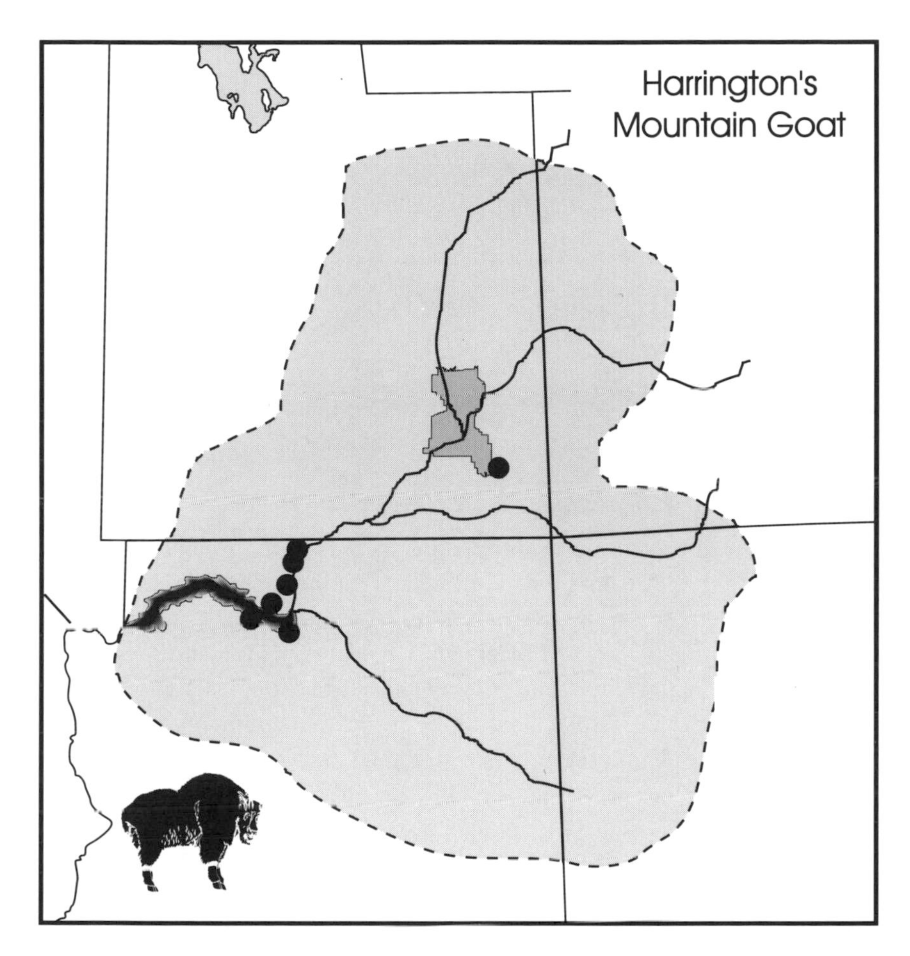

Figure 5.11. Map of Harrington's mountain goat fossil sites on the Colorado Plateau. (After Nelson, 1990.)

Bighorn sheep scale the canyon walls of Canyonlands today. They are reasonably abundant (and have even been suggested as a sort of mascot for the park), but they are not the most impressive members of their family to have roamed these canyons. During the late Pleistocene, an animal named Harrington's mountain goat (Fig. 5.5) lived throughout the Colorado Plateau. Fossils of this species have been found in rockshelters from the Grand Canyon, Glen Canyon, and Natural Bridges National Monument (Fig. 5.11).

This mountain goat left behind not only fossil bones but also hair and dung, preserved in dry rockshelters and caves. From these remains, Jim Mead and his colleagues at Northern Arizona University have pieced together the following portrait. Harrington's mountain goat was slightly smaller than its modern relatives (an unusual circumstance, since most Pleistocene megafaunal mammals were larger than their modern counterparts). Weighing from 60 to 80 kg (132–176 lb), it was stockier than the modern mountain goat, and it had the same white fur. It lived along the rocky slopes and canyon walls of the Colorado River and ranged down to the river's edge to graze on the vegetation there. Pollen extracted from ancient dung pellets in caves indicates that these animals bedded down there during the spring months, and possibly in late winter and early summer as well (based on the times of year that the plants in question produce their pollen). Amazingly, one cave in the Grand Canyon has preserved indentations from where Harrington's mountain goats laid, thousands of years ago. Like Bechan Cave, this cave also contained a thick mat of dung pellets, presumably deposited over many goat generations. Little Pleistocene time capsules like this are scattered throughout the canyon country of the Colorado Plateau. Who knows what we will find in the hundreds of caves and rockshelters not yet explored?

Again, the fossil evidence provides precise, incontrovertible evidence of how these animals lived. We do not have to speculate what the goats ate, based on their modern relatives in the Southwest or elsewhere. We know exactly what they ate, and even at what time of the year they ate it! The plant remains in the dung pellets revealed that Harrington's mountain goats fed on a variety of plants and that their diet shifted through the seasons, depending on the availability of various food plants. Among these plants were grasses, sedges, and rushes, in addition to smaller amounts of limber pine needles, globemallow, sagebrush, and yucca. Dung pellets from Natural Bridges National Monument also contained remains of Douglas-fir and joint fir. These plants grew at lower elevations on the Colorado Plateau during the Pinedale Glaciation than they do today.

Paleontologists working in other parts of the world can only dream of obtaining the kind of information on past plant and animal life that is readily available in the American Southwest. Our warm, dry climates and abundance of caves and rockshelters provide unmatched opportunities to study the ancient past. Although decomposition of soft animal and plant tissues is the rule for most of the world (and thank goodness for that, otherwise the world would be knee-deep in the refuse of ancient life!), it is the exception in dry caves.

The dung pellets offer one additional piece of information about this formerly widespread mountain goat species. The youngest radiocarbon age, based on samples of dung, is about 11,600 yr B.P. This age may be taken as a reasonable estimate of the time when they became extinct.

Ground sloths were also quite abundant on the Colorado Plateau during the last glaciation. The region had two species of ground sloths, Harlan's ground sloth and the Shasta ground sloth (Fig. 5.5). Their modern relatives are much smaller and live mostly in trees. But ground sloths, as their name implies, were ground-dwelling animals. The smaller of the two species was the Shasta ground sloth; it was about 2.75 m (9 ft) in length and may have weighed as much as 250 kg (550 lb). We have a good idea of this animal's appearance, because an intact, mummified specimen was found in Aden Crater, New Mexico. The Shasta ground sloth browsed on a variety of desert plants. The configuration of bones in its skeleton suggests that it waddled as it walked on the padded outer margins of its hind feet. Those hind feet were rather large and clumsy, with four giant claws. It probably could not run very far or very fast, but its slender front feet helped it collect a wide variety of food plants, so it did not have to travel far to feed itself every day. It used the claws on its front feet to rake the fruits off cactus pads, thus avoiding painful contact with cactus spines. It deposited dung pellets the size of baseballs. These were preserved in dry caves. The dung yielded the remains of globemallow, Mormon tea, yucca, century plant, various cacti, catclaw acacia, salt brush, and mesquite. There is no evidence of human predation on this ground sloth.

The other ground sloth, Harlan's ground sloth, was a true giant. Standing as much as 3.5 m (12 ft) tall, this sloth weighed perhaps six times as much as the Shasta ground sloth. It was a formidable beast; its leathery hide was reinforced with an outer layer of boney plates, called *dermal ossicles*. These served as a suit of armor, making it nearly invulnerable to attack. In fact, it appears that this animal, once grown, had no predators on the Colorado Plateau until the arrival of Paleoindians.

The great height of Harlan's ground sloth allowed it to browse from the limbs of trees. It pulled the limbs near its mouth with huge, powerful claws, then it pulled the leaves from the limbs by wrapping its large tongue around them, as it leaned back on a large, muscular tail. The two sloth species on the plateau were ecologically compatible (i.e., they did not compete for the same resources), because their size and feeding habits ensured that the food sources of the plateau region were partitioned between them: the Shasta ground sloth fed on plants near the ground while Harlan's ground sloth fed on the tree limbs above.

Both creatures became extinct about 11,000 yr B.P. Fossil specimens dating to the last glaciation have been found in several localities on the Colorado Plateau (Fig. 5.12). Sloths are some of the strangest of Pleistocene mammals; they have no modern counterparts anywhere in the world. Their only living relatives are relatively small creatures that spend nearly all of their time hanging from tree limbs, browsing on leaves. These slow-moving, shy creatures provide little evidence of the manner of life of their ancient relatives. For instance, at more than 700 lb and 12 ft, Harlan's ground sloth may have been slow moving, but it certainly had no reason to be shy; in fact it probably was not afraid of any other animal.

Next in our Pleistocene menagerie from the Colorado Plateau are two species of camels. Whereas the modern distribution of camels (in Asia and Africa) is strictly

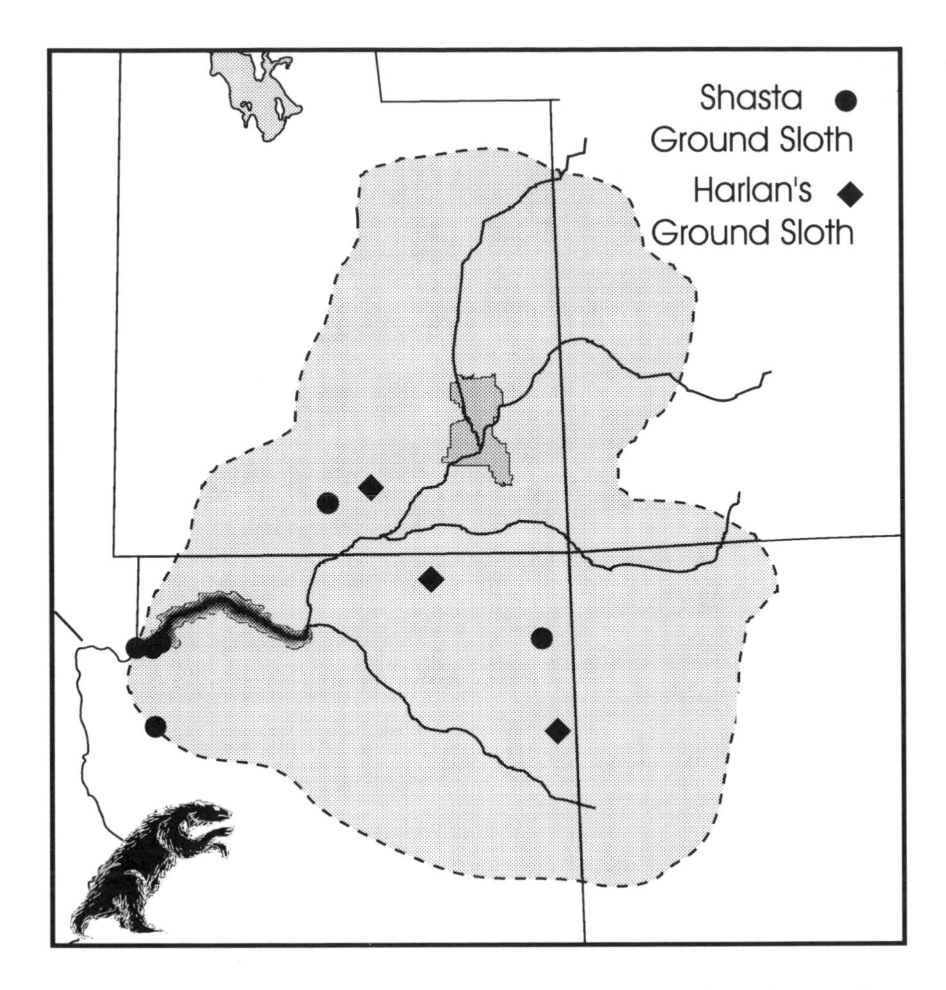

Figure 5.12. Map of ground sloth fossil sites on the Colorado Plateau. (After Nelson, 1990.)

Old World, the fossil record indicates that camels evolved in the New World, then spread to Eurasia. There are still members of the camel family in the New World. These include the llama, guanaco, and alpaca of the Andes Mountains and the Patagonia region of South America. During the Pleistocene, more species of camels were in existence. On the Colorado Plateau, these included Yesterday's camel (Fig. 5.5) and the giant Nebraska camel. The giant Nebraska camel, 4.2 m (14 ft) tall at the shoulder, was rare in this region, as far as we can tell from its scant fossil record there (Fig. 5.13). Yesterday's camel was one of the most abundant grazing animals of western North America, including localities in the Canyonlands region. It was about 20% larger than the modern dromedary of Africa. Dung pellets from this camel have been found in a cave on the plateau. Unlike its modern relatives in

Eurasia, Yesterday's camel was not a desert dweller but an inhabitant of grasslands, where it lived in herds. Accordingly, Yesterday's camel did not have a large hump on its back, an adaptation of its desert-dwelling relatives. Both of the Pleistocene camels became extinct about 11,000 yr B.P. The U.S. Cavalry brought modern camels into the American Southwest before the Civil War. The animals did well, but this experiment in desert transportation was short-lived.

Other grazers on the Colorado Plateau during the Pleistocene were ancestral horses (Fig. 5.5). Like the camels, horses are also thought to have evolved on the plains of North America. Pleistocene horses grazed the grasslands of this continent until the end of the Pleistocene, then became extinct. By then, their relatives had become established in the Old World, where they were domesticated; they were

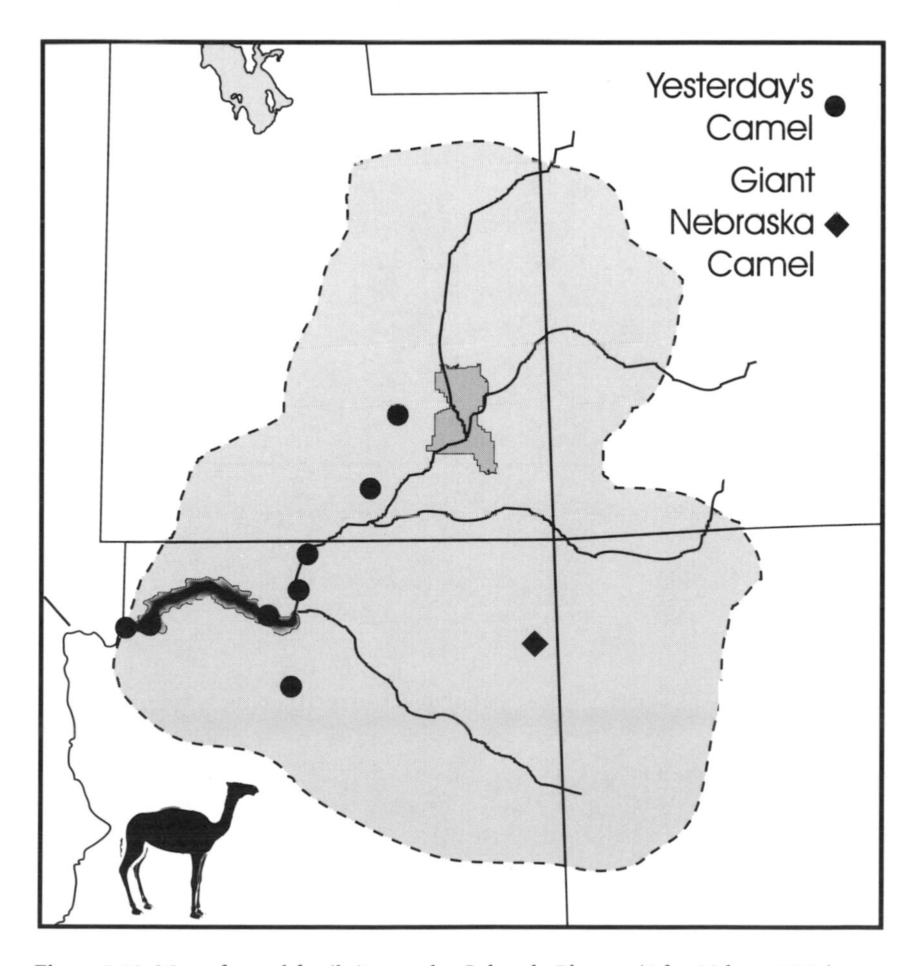

Figure 5.13. Map of camel fossil sites on the Colorado Plateau. (After Nelson, 1990.)

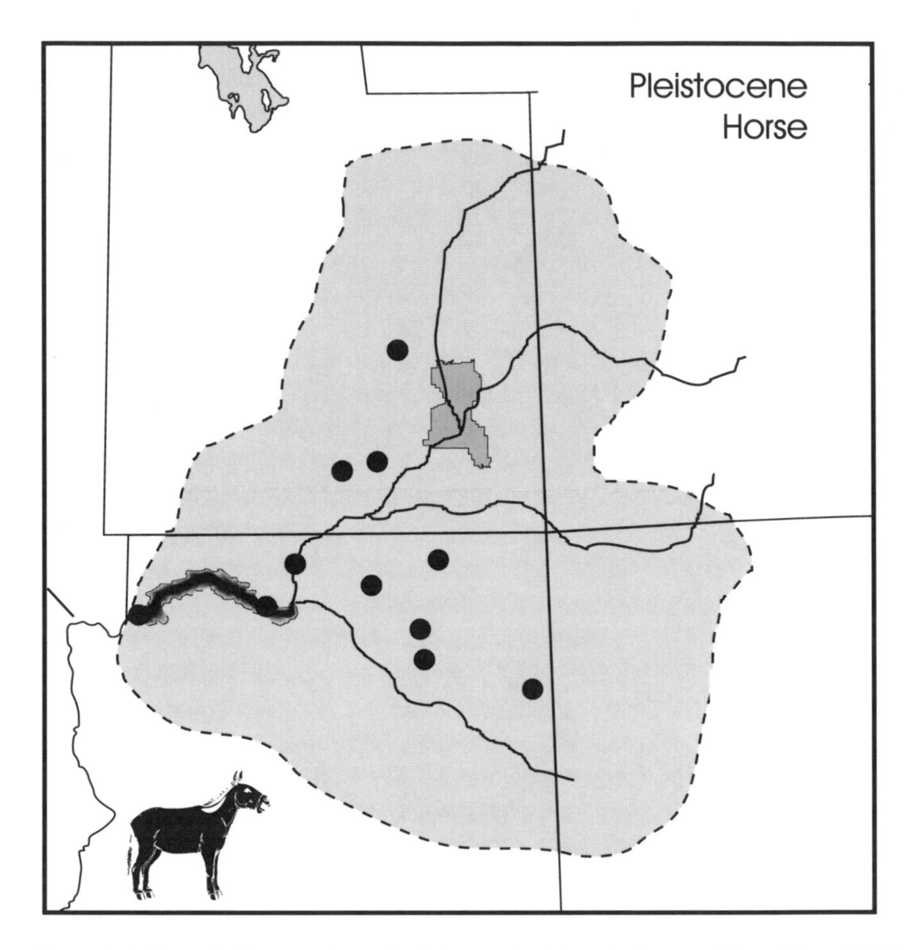

Figure 5.14. Map of Pleistocene horse fossil sites on the Colorado Plateau. (After Nelson, 1990.)

eventually brought back to the New World with European explorers and colonists. A common image of American Indians, reinforced in books and on film, is of the proud warrior on horseback, hunting buffalo or loosing arrows in remarkable feats of equestrian combat. In fact, Indians had to make do without modern horses until the sixteenth century, when horses were introduced by the Spaniards. One of the more amazing facts about prehistoric Indians in the New World is what they managed to accomplish without such animals as the horse and ox. The largest draft animal used by Indians prior to European contact was the dog, who was used to pull relatively light loads on sledges and travois. Yet among their list of architectural accomplishments are the Mayan and Aztec temples of Mexico and Central America, the city of Machu Pichu high in the Andes of Peru, Pueblo Bonito in Chaco Canyon, and Cliff Palace at Mesa Verde, all built solely by human labor.

During the late Pleistocene, as many as three species of ancient horses lived on the Colorado Plateau (Fig. 5.14). These included ancestral forms of the ass, a burrolike horse, and a larger species comparable to the modern quarter horse. Paleontologists have not reached a consensus about the number or names of Pleistocene horse species in North America.

Another group of Pleistocene mammals of the Colorado Plateau were the bovine species, the shrub ox and the musk-ox. As I mentioned earlier, musk-ox remains have been found as far south on the Colorado Plateau as the Dolores River of southern Colorado (Fig. 5.15). Another species of musk-ox on the Colorado Plateau, the bonnet-headed musk-ox, belonged to a different genus than the modern arctic musk-ox.

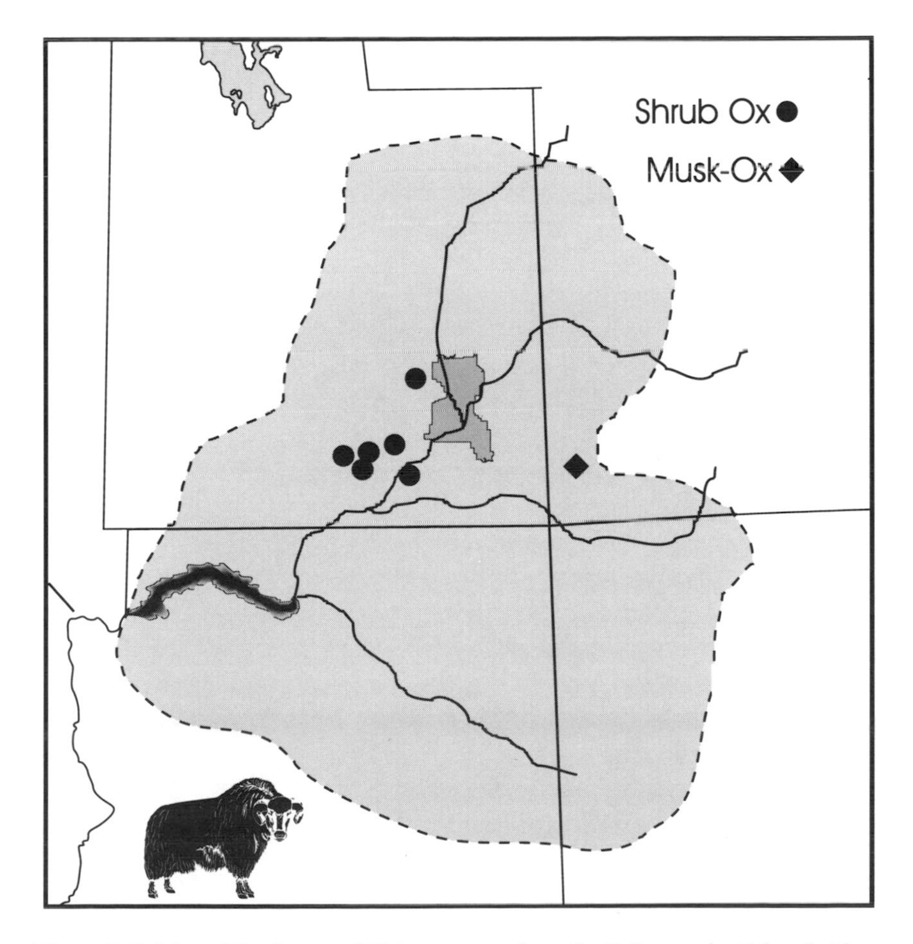

Figure 5.15. Map of shrub-ox and Pleistocene musk-ox fossil sites on the Colorado Plateau. (After Nelson, 1990.)

The shrub ox was apparently more widely distributed on the Colorado Plateau and elsewhere (Fig. 5.15). It ranged from northern California to central Mexico and as far east as Illinois. Based on its skeleton, it was more closely related to modern cattle, which originated in Asia. It had recurved horns, not unlike the shape of Asian goat horns. Cave deposits from the plateau have yielded dung, hair, and teeth of this animal. Remains of shrub oxen were found in the same stratigraphic horizon as human artifacts in Burnet Cave, New Mexico. The only member of the bovine family that survived the transition from the last ice age into the Holocene was the American bison.

Thus was the Colorado Plateau populated with herds of large grazing and browsing mammals during the late Pleistocene. It was an American Serengeti. The Serengeti Plain of Tanzania is perched on a plateau over 5000 ft (1500 m) in elevation. Annual precipitation ranges from 500 to more than 750 mm (20 to more than 30 in.). The savannah vegetation of this plain feeds many grazing and browsing animals, including antelope, zebra, elephant, hippopotamus, rhinoceros, and giraffe. The Colorado Plateau probably received similar amounts of precipitation in the late Pleistocene. At that time, there were several counterparts to the Serengeti fauna. Instead of African antelope, there were pronghorn antelope; in place of zebra, there were Pleistocene horses. The role of the elephant was played by mammoths and mastodons. If you combine the tough hide and sheer bulk of a rhinoceros with the height of a giraffe, you arrive at something like Harlan's ground sloth. The big difference is that the grazers of the Serengeti have survived the Pleistocene-Holocene transition, whereas nearly all of the Pleistocene grazers of the Colorado Plateau became extinct.

The Pleistocene predators that roamed the Colorado Plateau were likewise a formidable group, including wolves and coyotes, large cats, and bears (Fig. 5.5). Again, a comparison with the Serengeti Plain fauna is not unreasonable. In place of the African lion, there was the North American lion. In place of hyenas and jackals, there were wolves, dire wolves, and coyotes. These predators made the Colorado Plateau a dangerous place for a hoofed animal to take a nap, give birth to young, or wander away from the herd.

In the canine family, the dire wolf was perhaps the most interesting. This animal was considerably larger than its modern relatives, with powerful jaws for attacking large prey animals. It had sturdier limbs and a larger, broader head than the modern wolf. Its teeth were more powerful than those of any living member of the canine family; it was undoubtedly a dangerous predator, and it also scavenged carcasses, like its modern relatives, wolves and coyotes. Large numbers of dire wolf skulls are on display at the Page Museum in Los Angeles, California. Those wolves were attracted to dead and dying animals associated with tar seeps, such as the La Brea tar pit, where thousands of animal bones were preserved in layers of sticky asphalt. Dire wolves are thought to have roamed the western hemisphere from southern Alberta, Canada, to Peru. The dire wolf became extinct by 9400 yr B.P.

The large cats that hunted on the plateau during the last ice age include the saber-toothed cat, the American lion, and the mountain lion. The jaguar may also have lived there, although its closest fossil record from the late Pleistocene is at Dry Cave in southeastern New Mexico. There were several species of saber-toothed cats in the New World during the Pleistocene. Their large, daggerlike incisors were finely serrated to further enhance their lethality. What was the purpose of the saber teeth? The big, slashing incisors undoubtedly proved useful for piercing the thick, tough hides of large ice-age herbivores. The first stabbing bite had to count, because megafaunal mammal hunting was a dangerous business.

To the untrained eye, an attack by a large cat on a scampering hoofed animal seems dangerous only to the intended victim, not to the predator. But biologists have observed that the hunter takes almost as many risks as the hunted. Here is why: the predator has to be in excellent physical condition to pursue its prey, especially prey species that elude predation by their speed of escape. The predator gets as close as it can to the prey animal before it begins the chase, because if the chase goes on too long, the predator uses up too much energy, and the eventual meal that it gets may not make up for the calories expended in the chase. If a predator is injured during the attack, it may never again be in sufficiently good shape to hunt. Such injuries often are the result of kicks from hoofed prey animals. Although these kicks may not kill a predator immediately, they can break ribs, legs, or jaw bones. A large cat with a broken jaw is doomed to starve to death in short order. Likewise, a cat with a broken leg will never run as fast again, even if the break manages somehow to heal. Predators that have been injured while attacking prey are seldom swift enough to continue the chase. So it is vitally important for large cats, as solitary hunters, to make one, quick, decisive attack on a prey animal. In an age of huge, powerful, hoofed prey, saber teeth were a decided advantage.

Why were there no saber-toothed wolves or other members of the canine family? The answer probably lies in the style of hunting that dogs and their relatives use when pursuing hoofed animals. As we can see today when a pack of wolves attacks a herd of caribou, canine hunters hunt in groups, wearing their prey down with a coordinated series of darting movements, feints, sprints, and stalemates. In this hunting technique, the risk to the individual hunters is lessened. The prey animal, usually a sick, feeble, or juvenile individual culled from the herd, becomes exhausted from the relentless pursuit of several animals; the pack animals take turns in the chase, and by the time they close in for the kill, the prey animal is often too weak to put up a fight. The debility of the prey is another safety factor for pack-hunting predators. Thus their first bite on the prey animal does not count for as much as the first bite attempted by a large cat on the same prey animal.

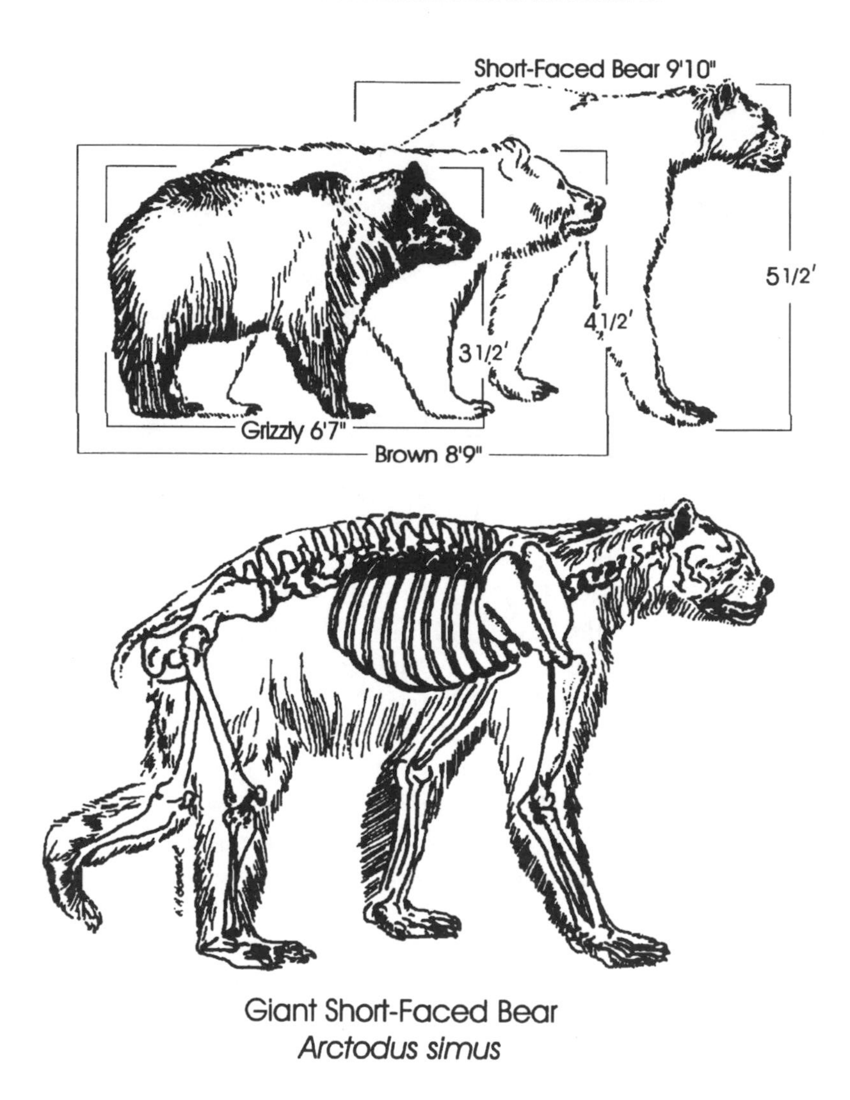

Figure 5.16. Size comparison between the Pleistocene short-faced bear and the modern grizzly bear. (After Border, 1988.)

The Pleistocene lion that hunted throughout much of the northern hemisphere probably also hunted in groups or *prides.* It was quite a bit larger than the modern lion but otherwise looked quite similar to it (Fig. 5.5). It is believed that the Pleistocene lion had a tuft of hair on its tail, based on cave paintings in Spain (paleontologists take their clues wherever and however they can find them). At Jaguar Cave in Idaho, bones of the Pleistocene lion have been found in the same level as hearths dated at 10,400 yr B.P., but there is no evidence that humans killed and butchered the lions (i.e., no butchering cut marks were found on the bones).

Grizzly and black bears were a part of the ice-age fauna of the Colorado Plateau, but their extinct relative, the short-faced bear, was decidedly more terrifying than even the largest grizzly bear. The short-faced bear has been called the most powerful predator of the American Pleistocene. It was a long-legged, short-bodied animal, with a short face and a broad muzzle. Standing 2 m (6 ft) tall at the shoulder, it was 3 m (10 ft) long from nose to stubby tail. It dwarfed its modern relative, the grizzly bear (Fig. 5.16).

The long legs of the short-faced bear were adapted for running down prey animals. Modern grizzlies, with their relatively short legs, can sprint at 65 km (40 miles) per hour over short distances. The short-faced bear certainly ran even faster and was probably able to sustain high speeds over much greater distances. Its massive jaws with their large upper incisors were powerful weapons with which to attack prey animals (Fig. 5.17). We have no evidence that short-faced bears preyed on humans, but some archaeologists have suggested, only half jokingly, that North America was not a safe place for people to live until the short-faced bear died out, about 11,000 years ago.

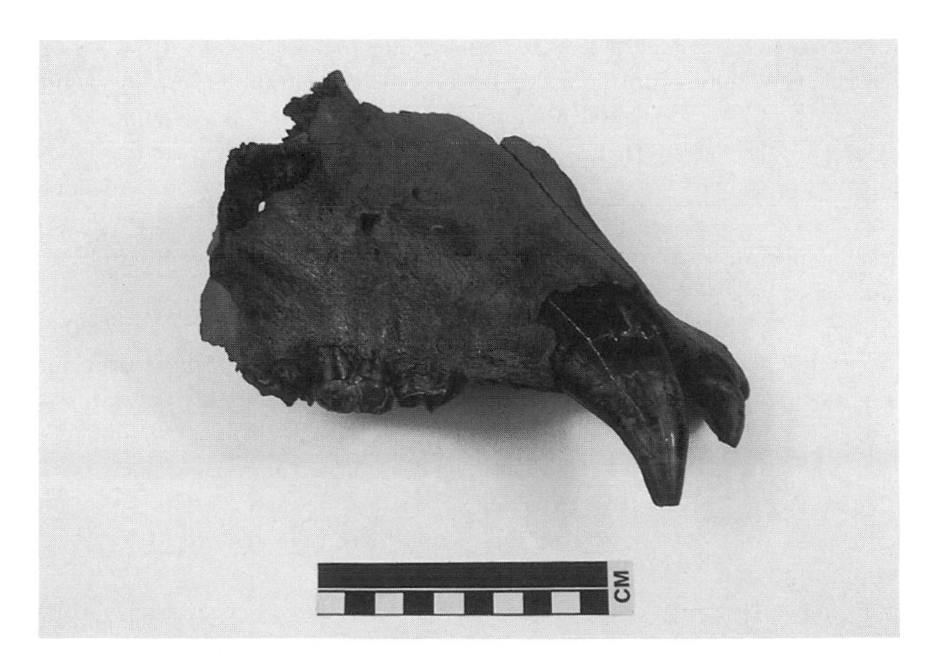

Figure 5.17. Lateral view of the upper jaw of a short-faced bear from Huntington Canyon, Utah. (Photograph by David Madsen, courtesy Utah Geological Survey.)

Grizzly bears are thought to have emigrated to the New World from Asia via the Bering Land Bridge sometime late in the Pleistocene. Remains of the short-faced bear and grizzly bear have been found together in a cave in Wyoming, so the two species shared the land for at least a few centuries or millennia before the short-faced bear died out. Short-faced bear remains were also found in late Pinedale–age sediments at Huntington Canyon, Utah.

Early Peoples of the Canyonlands Region

The canyons cut by the Colorado and Green rivers contain many small cliff dwellings, some perched high on canyon walls. Canyonlands National Park was home to the Anasazi people, although it was near the northwestern boundary of their culture. Rock art, including both **petroglyphs** and **pictographs** (Fig. 5.18), abounds in this region.

A chronology of ancient cultures of the American Southwest is shown in Table 5.1. The bulk of archaeological evidence points to the Clovis culture as the earliest group of Paleoindians to occupy North America. The oldest Clovis artifacts have been radiocarbon dated to just after 12,000 yr B.P. The style of **projectile points** that characterized Clovis culture only persisted for a few hundred years; then they were replaced by another type of point, characteristic of the Folsom culture. Both cultures are thought to have been big-game hunters. Archaeologists believe that Clovis and Folsom hunters were only infrequent visitors to the canyon country of the plateau. Their search for big game animals probably led them most often to wide-open plains regions, where such game animals were both more common and easier to attack. Clovis kill and camp sites were most often located near sources of water, such as small streams or springs. Perhaps Clovis hunters ambushed their prey as the animals approached these watering places.

Folsom peoples did pass through the Colorado Plateau, but proof for their occupation there has been hard to come by. Most evidence consists of individual Folsom points, found on the surface of the ground. No evidence of a Folsom camp was found on the plateau until 1975, and the first Folsom artifacts were found in southeastern Utah in 1984. Archaeologists have surveyed most of the Colorado Plateau, because of the great attraction of Anasazi sites and artifacts, so Folsom sites must truly be rare here or they would have been discovered in greater abundance.

Starting about 8000 yr B.P., people of the Archaic culture were probably the first to utilize the Canyonlands region widely, hunting smaller game animals and foraging for food plants. Archaic sites dot the map of the Colorado Plateau like holes on a well-used dartboard. Unfortunately, archaeological studies of the Ar-

Figure 5.18. Pictograph near Moab, Utah, possibly depicting a Pleistocene mammoth. (Photograph by David Madsen, courtesy Utah Geological Survey.)

Table 5.1. Chronology of the Development of Paleoindian, Archaic, and Anasazi Cultural Periods

Time Interval (yr B.P.)	Cultural Period	Regional Events
12,000–800	Paleoindian	Nomadic big-game hunting; Clovis and Folsom cultures
8000–2100	Archaic	Nomadic hunter-gatherer culture; expanded tool kit; numerous tool-making traditions develop
2100–1600 (100 B.C.–A.D. 400)	Basketmaker II	Shift from nomadic to sedentary life-style in caves and rockshelters; beginnings of agriculture (corn)
1600–1300 (A.D. 400–700)	Basketmaker III	Early ceramics; domestication of turkey; beans domesticated; building of pit houses
1300–1100 (A.D. 700–900)	Pueblo I	Aboveground dwellings built; further development of ceramics; development of kiva; cotton domesticated
11–900 (A.D. 900–1100)	Pueblo II	Anasazi reach their maximum distribution; small pueblos built; “Chaco Phenomenon” began
900–700 (A.D. 1100–1300)	Pueblo III	Height of Anasazi cultural development; cliff dwellings at Mesa Verde completed, then abandoned; “Chaco Phenomenon” comes to an end
700–400 (A.D. 1300–1600)	Pueblo IV	Fewer sites occupied; sites on Colorado Plateau and Mogollon highlands deserted by A.D. 1450; development of Hopi and Zuni communities

Figure 5.19. Newspaper Rock, Canyonlands National Park. (Photograph by the author.)

chaic Period have been few. Between the excitement of studying the earliest human cultures to invade this region (Clovis and Folsom peoples) and the overwhelming architectural wonders of the Anasazi culture, archaeologists have given little attention to the Archaic peoples and their way of life. In a sense, the Clovis and Folsom peoples had an easier time gathering enough food; they had a wide variety of megafaunal mammals to hunt. How did the hunters of the Archaic Period adapt to the loss of this major food resource? Somehow, they not only survived on the Colorado Plateau, they spread far and wide as their populations grew in the early Holocene, occupying every part of the plateau country. If Clovis and Folsom peoples were responsible for the destruction of the megafauna, then it can truly be said that the Archaic peoples were the first inhabitants of this region to live in "harmony" with their environment (they did not destroy the natural resources of the region through more than 8000 years of continuous occupation).

The best-preserved archaeology of the region relates to the Anasazi culture, however. Visitors from John Wesley Powell onward observed extensive Anasazi ruins and rock art in Canyonlands and Glen Canyon (Figs. 5.19 and 5.20). Archaeological reconnaissance began in earnest with the Hayden Survey in 1873–1876. Some of the Wetherill brothers, from the same family that discovered Mesa Verde, were the first to excavate ruins at Moqui Canyon, a tributary of Glen Canyon. Lengthy studies made by Byron Cummings and his students from the University of

Figure 5.20. Cliff dwelling perched on a canyon wall along the Colorado River, illustrated for John Wesley Powell's book, *Canyons of the Colorado.*

Figure 5.21. The Needles region of Canyonlands National Park. (Photograph courtesy of Corel Corporation.)

Utah during the first two decades of this century, then by Donald Scott, Noel Morse, and Henry Roberts in the 1920s and 1930s, established that another culture shared this region with the Anasazi. These were the Fremont people, hunter-gatherers whose domain stretched to the north of the Colorado River in Utah. Anasazi ruins representing the Basketmaker II and III periods, as well as the Pueblo I, II, and III periods, have been found throughout southeastern Utah. These cultures are discussed in more detail in Chapter 7.

When Congress authorized the construction of Glen Canyon dam in 1956, an archaeological survey of the region was undertaken. The researchers had to work fast, because many sites would soon be under water. The reservoir, Lake Powell, now covers 295 km (183 miles) of the Colorado River course, with 2900 km (1800 miles) of shoreline. Along with occupation sites, the researchers documented enormous rock art, depicting human figures with elongated or trapezoidal bodies, along with various animals and geometric designs. Excellent examples of rock art can be seen in the Needles and Maze regions of Canyonlands (Fig. 5.1). More than 700 archaeological sites have been documented in the Needles region alone (Fig. 5.21). More can be found in the Horseshoe Canyon region, just west of the park.

Canyonlands and the adjacent Glen Canyon Recreation Area represent some of the most rugged, beautiful canyon country in the world. It is a region with a fascinating history, along with the rest of southeastern Utah. These canyons and plateaus were home to a diverse megafauna in the Pleistocene. Scientists are not sure why they became extinct at the end of the last ice age; I will pursue this topic more fully in the next chapter. This region represented a boundary zone between Anasazi and Fremont peoples near the end of prehistory. It remains a land that is not easily traversed, but that is much admired for its beauty, its pristine wilderness, and the grandeur of its twisting, turning canyons.

Suggested Reading

Agenbroad, L. D. 1990. *Before the Anasazi: Early Man on the Colorado Plateau.* Flagstaff: Museum of Northern Arizona. 32 pp.

Betancourt, J. L. 1990. Late Quaternary biogeography of the Colorado Plateau. In Betancourt, J. L., Van Devender, T. R., and Martin, P. S. (eds.), *Packrat Middens: The Last 40,000 Years of Biotic Change.* Tucson: University of Arizona Press, pp. 259–293.

Cordell, L. S. 1984. *Prehistory of the Southwest.* New York: Academic Press. 409 pp.

Frison, G. C. 1991. *Prehistoric Hunters of the High Plains,* Second Edition. New York: Academic Press. 532 pp.

Lister, R. H., and Lister, F. C. 1983. *Those Who Came Before.* Globe, Arizona: Southwest Parks and Monuments Association. 184 pp.

Mead, J. I., and Bell, C. J. 1994. Late Pleistocene and Holocene herpetofaunas of the Great Basin and Colorado Plateau. In Haroer, K. T., St. Clair, L. L., Thorne, K. H., and Hess, W. M. (eds.), *Natural History of the Colorado Plateau and Great Basin.* Boulder: University of Colorado Press, pp. 255–275.

Mead, J. I., Agenbroad, L. D., Davis, O. K., and Martin, P. S. 1986. Dung of *Mammuthus* in the arid southwest, North America. *Quaternary Research* 25:121–127.

Mead, J. I., O'Rourke, M. K., and Foppe, T. M. 1986. Dung and diet of the extinct Harrington's mountain goat (*Oreamnos harringtoni*). *Journal of Mammalogy* 67:284–293.

Nelson, L. W. 1988. *Mammoth Graveyard: A Treasure Trove of Clues to the Past.* Hot Springs, South Dakota: The Mammoth Site. 24 pp.

Nelson, L. W. 1990. *Ice Age Mammals of the Colorado Plateau.* Flagstaff: Northern Arizona University Press. 24 pp.

Powell, J. W. 1895. *Canyons of the Colorado.* Reprinted in 1961 under the title *The Exploration of the Colorado River and Its Canyons.* New York: Dover Press. 400 pp.

Withers, K., and Mead, J. I. 1993. Late Quaternary vegetation and climate in the Escalante River basin on the central Colorado Plateau. *Great Basin Naturalist* 53:45–61.

6

GRAND CANYON NATIONAL PARK
A Voyage through Time

Grand Canyon National Park contains the Colorado River canyon, several side canyons cut by smaller streams, and adjacent plateau regions of the north and south rims (Fig. 6.1). According to current geological theory, the rocks visible at the bottom of the Grand Canyon, a thousand meters (3300 ft) or more below the rim, were formed during Pre-Cambrian times, more than a billion years ago. Tourists peer into the canyon and are overwhelmed with the immensity of its depths (Fig. 6.2). Paleontologists and geologists are equally awed by the immensity of the geologic record encompassed in the same view.

Surprisingly, the canyon itself is far younger than the rocks exposed along its walls. Geologists believe that most of the cutting of the canyon by the Colorado River has taken place in the last 6 million years. For more than 200 million years after the last rock layers of the canyon were deposited, the region was a relatively undisturbed plateau.

The Grand Canyon is a place that simply must be seen to be appreciated. But when you are there, do not let the staggering proportions of the landscape blind you to the fact that the canyon is a vital, living place, not just an enormous tomb for very old fossils or a sterile cavern of barren rock. Much has happened in and around the Grand Canyon during the late Pleistocene and Holocene. Some of the

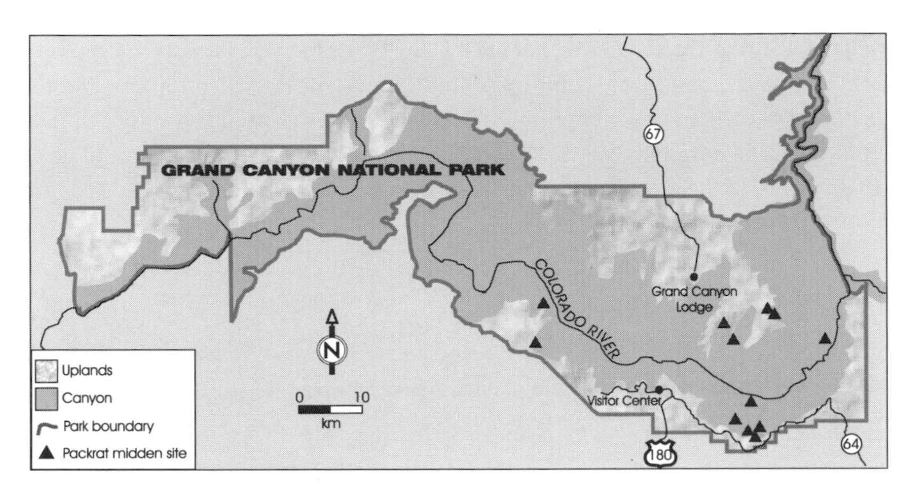

Figure 6.1. Map of Grand Canyon National Park, showing sites discussed in the text.

Figure 6.2. The Grand Canyon from the south rim. (Photograph by the author.)

most interesting creatures lived here just a few thousand years ago. If the geologic history of the canyon were compressed into one 24-hr day, then the late Pleistocene period I will be describing would pass by during the last 2 seconds of the last minute of that day (11:59:58 and 11:59:59 P.M.). But during that geologic blink of an eye, some of the most fascinating fossils were preserved in dry caves tucked into the canyon's walls.

Modern Setting

The walls of the Grand Canyon span more life zones than are encompassed within any other national park in the United States (Fig. 6.3). The floor of the Grand Canyon drops 600 m (2000 ft) along the length of the canyon, from an elevation of about 900 m (3000 ft) above sea level at the upstream end of the canyon to an elevation of about 300 m (1000 ft) at the downstream end; the north rim of the canyon averages about 2700 m (9000 ft), and the south rim is about 2100 m (6900 ft). This is a vertical relief of more than 2000 m (6600 ft). A raven soaring down the canyon from the north rim to the river (a feat frequently performed by ravens, especially during the mating season) would travel from the boreal woodland zone on top to the Mojave Desert zone at the bottom. Were this trip through the ecosystems to be taken across the latitudes rather than down a canyon, the raven would have to begin at Val d'Or, Quebec and fly to Las Vegas, Nevada.

The modern vegetation of the canyon does not follow strict elevational zones. This is because soil moisture and precipitation vary so much within the diverse environments of the canyon. As elsewhere, the principal division between arid and moist regions falls along a north-south gradient. The north-facing side of the canyon would naturally be cooler and moister than the side facing south, but at the Grand Canyon the north side is also considerably higher in elevation, so it receives even more moisture. This becomes clear to any visitor who arrives in winter. The south rim remains open all year, but the north rim is closed until late spring, because the access road to the north rim crosses the high Kaibab Plateau, buried in deep snow. These conditions also existed during the late Quaternary.

In moister localities, piñon-juniper woodlands extend down to about 1350 m (4430 ft); in drier localities, the lower boundary of this woodland falls at about 1550 m (5090 ft). Conifer forests with firs and ponderosa pine range down to about 1900 m (6230 ft) in moist areas, but only down to about 2450 m (8040 ft) in dry areas.

The north rim of the canyon is the high point of the hump-shaped Kaibab Plateau. This region receives up to 660 mm (26 in.) of precipitation per year. The south rim receives only 400 mm (16 in.) of precipitation per year. The life zones of

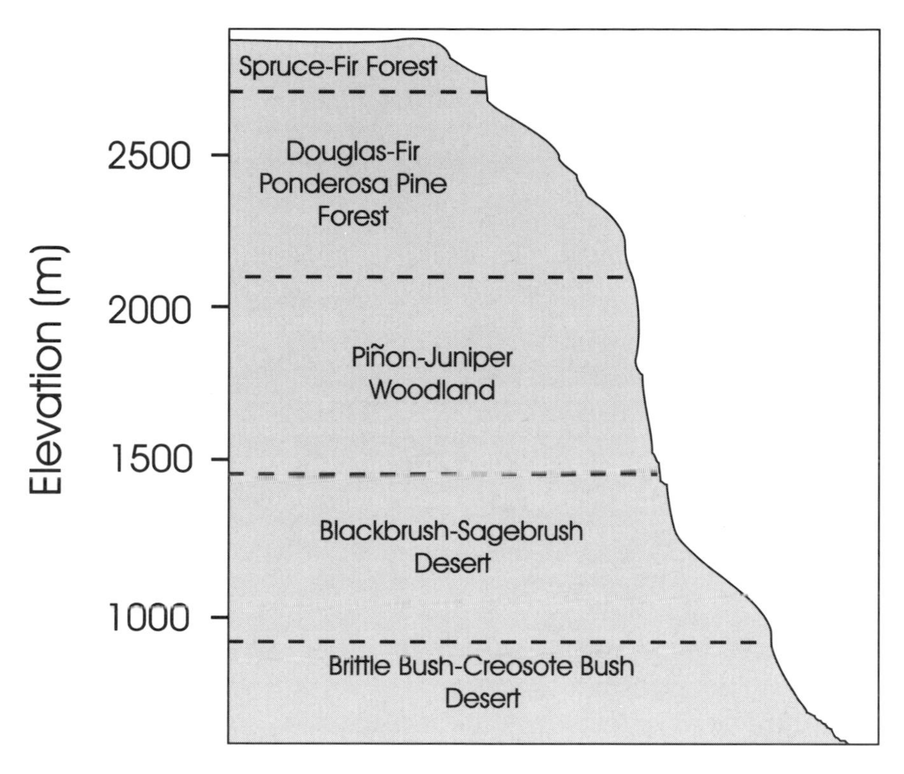

Figure 6.3. Life zones of the Grand Canyon.

the canyon are not sharply defined, and wildlife moves freely between them. The spruce-fir forests and meadows that grow on the plateau country of the north rim are home to abundant mule deer populations. Interspersed in the conifer forest are stands of aspen and meadows with beautiful wildflowers, including wild orchids, wild strawberry, columbine, larkspur, and white geranium. The conifers provide food (bark and twigs) for porcupines and the Kaibab squirrel, a black squirrel with a white tail that is found nowhere else. Close to the north rim, a drier, more open forest dominated by ponderosa pine resembles that of the south rim. Shrubby plants in this region include Gambel oak, mountain mahogany, locust, snowberry, and elderberry. A different species of squirrel lives in the ponderosa pine forests of the south rim. The tassel-eared Abert squirrel lives in the tall trees, where it builds nests in the tall crowns and feeds mostly on ponderosa pine seeds and twigs. It is thought that the two squirrel species diverged during the last glaciation when their ranges became separated.

There is abundant life on the plateau country surrounding the canyon, but regional species must work harder to find a foothold on the canyon walls themselves. Piñon pines set their roots into cracks and crevices, and Utah junipers cling tenaciously to narrow ledges lower down the canyon. Talus slopes provide a short-lived surface that must be recolonized after each new rock slide. Until quite recently, the places where sandy riverbanks formed along the Colorado River were just as ephemeral, as spring floods scoured the banks back to bedrock in some places and deposited a new load of gravel, sand, and mud in others. This annual riverbank disturbance has been greatly altered by the construction of the Glen Canyon Dam, upstream from the Grand Canyon. The previous spring torrents are now being caught in a reservoir, to be released downstream when the Bureau of Reclamation sees fit. However, plenty of good desert habitat remains at and near the bottom of the canyon, especially along the side canyons. The air that lies along the floor of the canyon may be almost 20° C (36°F) hotter than the air on the north rim. Most of the rain that falls into the canyon evaporates before it reaches the bottom; only the river supplies water to this arid region.

Away from the water's edge, the vegetation is tough and thorny. Catclaw acacia, cholla cacti, brittle bush, and creosote bush dot the rocky desert landscape. This region is home to the Grand Canyon rattlesnake, whose pink scales blend in with the rocks at the bottom of the canyon. Another desert denizen in residence here is the chuckwalla. This large, heavy-bodied lizard is related to the Gila monster, but the chuckwalla lacks the poisonous venom of its relative. Gila monsters also occur in the lower parts of the Grand Canyon. The chuckwalla is extremely warm adapted. It hardly moves around until the temperature reaches 38°C (100°F). The northern part of the canyon has biological affinities with the Great Basin; the southwestern end has closer affinities with the Mojave Desert, as represented by such plants as creosote bush, white bur sage, and ocotillo. These species probably became established in the canyon after the end of the last ice age, migrating in from the western end of the canyon, where the edge of the Mojave Desert region comes closest to the river.

Where smaller streams feed into the river, quieter water prevails, and with it comes lush vegetation (Fig. 6.4). Pools of water host mallard, merganser, and goldeneye ducks; herons; and other waterfowl. Osprey patrol the water, hunting for fish. Golden eagles circle down from nests higher up on the canyon walls. The greatest of the side canyons is Havasu, home to the last remaining native peoples of the Grand Canyon. Their village is about 16 km (10 miles) south of the river, by Havasu Creek. Their name is the *Havasupai,* or "people of the blue-green water." There are massive **travertine** deposits here that probably accumulated in the Pleistocene. Aprons of travertine line the banks above the warm, shallow water.

Figure 6.4. Havasu Creek. Note the lush vegetation in this sheltered location. (Photograph courtesy of Corel Corporation.)

Late Pleistocene Vegetation History

The complexities of the modern vegetation zones of the Grand Canyon were also present in the last glaciation. Just as today, the principal environmental gradient in the late Pleistocene divided the north rim (cool and moist) from the south rim (warm and dry). Accordingly, vegetation zones dipped farther down the canyon on the north side than on the south (Fig. 6.5). National Biological Service paleobotanist Ken Cole worked out the history of late Pleistocene vegetation in the canyon, based on plant remains from packrat middens. Fortunately, there is a rich midden record from this region, as the many rockshelters and caves in the canyon's walls provide excellent nesting sites for packrats. Cole studied the fossil record preserved in 52 packrat middens from the eastern Grand Canyon. His samples range in age from 24,000 yr B.P. to recent.

During late Pinedale times, modern vegetation zones lowered by 800–1000 m (2600–3300 ft). However, as might be expected, late Pinedale vegetation was not merely an exact replica of the modern vegetation, displaced downslope. Rather, different mixtures of species came together in the last glaciation, then went their separate ways when the glacial period was over. The main plant associations during

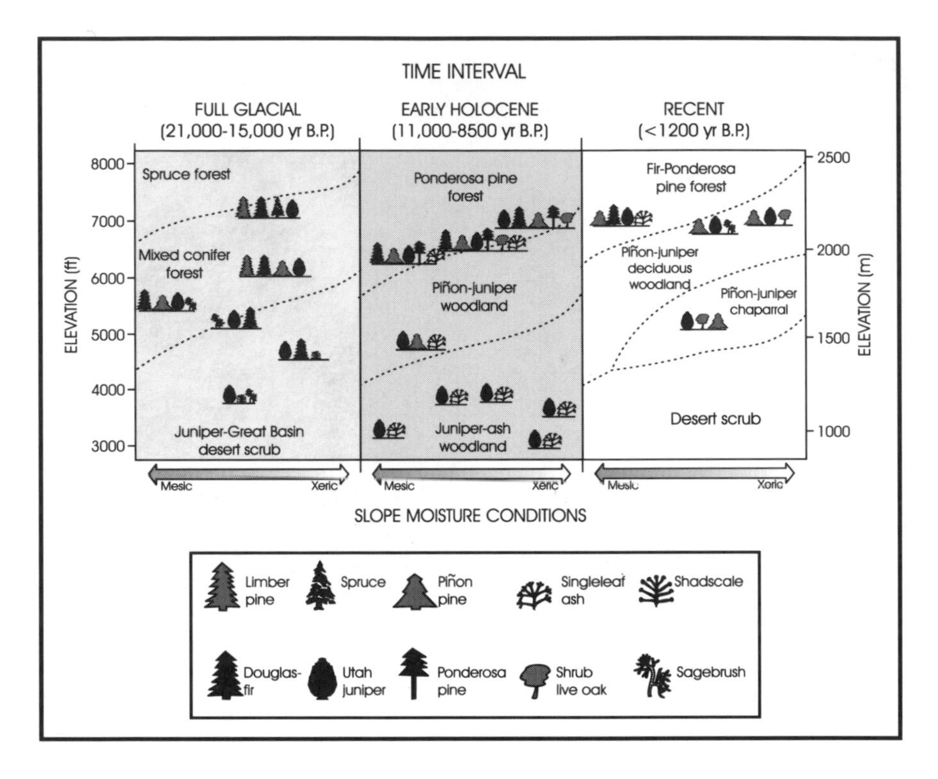

Figure 6.5. Illustration of the shifts in elevation of major plant macrofossil assemblages through the last 21,000 years in the Grand Canyon. (After Cole, 1990.)

the late Pinedale were juniper–desert scrub and such conifers as limber pine and spruce. Many hot-desert species—such as creosote bush, brittle bush, ocotillo, and white bursage—were not found in the Grand Canyon during the last glaciation. The lower (downstream) end of the Grand Canyon lies in the southeastern corner of the Mojave Desert today. This region supported more succulent species in the late Pinedale, including barrel cacti and yuccas.

Fossil plant remains of juniper–desert scrub vegetation have been found in packrat middens ranging in elevation from 450 to 1450 m (1475–4750 ft) in the Grand Canyon. The juniper was Utah juniper, the dominant juniper species in modern piñon-juniper woodlands across the Colorado Plateau region. The desert scrub species in the late Pinedale associations include Mojave prickly pear cactus, wild rose, sagebrush, and Utah agave. In middens below 1200 m (3900 ft), shadscale was particularly abundant. Below 700 m (2300 ft), blackbrush and snowberry remains are well represented, especially in middens from the western Grand Canyon.

The mixed conifer forest that developed in the late Pinedale occupied landscapes ranging from 1450 to 2200 m elevation (4750–7200 ft). The lower parts of this range were dominated by Douglas-fir and limber pine. With increasing elevation, Englemann spruce and white fir became more important. Spruce was the dominant tree above 2200 m on the north rim. These high forests also contained common juniper and a shrub called "mountain lover." These plants are common understory species in modern subalpine spruce-fir forests in the region. Of the Pinedale conifers that grew in and around the canyon, only white fir and Douglas-fir remain in sufficient numbers to ensure a reasonable chance of their being recorded in a modern packrat midden. Their modern elevational limits are 720–800 m (2350–2600 ft) higher than they were during the last glaciation. The other Pinedale conifer species are either very rare in the Grand Canyon (growing in isolated patches with unusual microenvironments) or missing altogether. However, similar associations of conifers can be found today farther north, in Utah.

The elevations of the boundaries between ecological communities, or *ecotones*, were approximately the same in the late Pleistocene as they are today, except that different communities occupy the various zones. The reason for this stability in the elevation of ecotonal boundaries is that they represent natural breaks in the topography of the landscape and in bedrock types. For instance, the ecotonal boundary at about 1450 m lies at the top of the Redwall Limestone formation in the canyon. Above this elevation there are gentle slopes, underlain by Supai Group shales. Below it lie steeper slopes and limestone cliffs.

One of the more intriguing absentees from the Pinedale midden record is piñon pine. This tree, so ubiquitous on the Colorado Plateau today, has only been found in one glacial-age midden sample from the western Grand Canyon. It was present in several late Pinedale middens in Wupatki National Monument, 65 km (40 miles) south of the Grand Canyon. These midden samples have been dated at about 11,500 yr B.P. Piñon pine may have grown in the western Grand Canyon during late Pinedale times, but apparently it was never abundant there during the last glaciation. Another piece of negative evidence concerning piñon pine in the Grand Canyon during the last glaciation is its absence from 30,000 yr B.P. onward at Stanton's Cave. The paleontology and archaeology of this cave are documented in a volume by Robert Euler. The eastern boundary of piñon pine apparently lay in the western or central Grand Canyon. Compared to the eastern reaches of the canyon, these regions are just low enough in elevation to have a warmer microclimate, allowing piñon pines to maintain a presence there through the end of the last glaciation. As I mention in Chapters 5 and 7, this tree was also absent from those regions during Pinedale times. At that time, it was restricted to warmer lowland regions to the south and west of the Colorado Plateau. It is hard to think of this region being practically devoid of piñon pine, because today there are literally millions of these trees in the piñon-juniper woodlands that dominate the region.

The history of limber pine is also quite interesting. This species had the most abundant fossil record of any Grand Canyon conifer during Pinedale times. Limber pine needles have been found in all Pinedale middens above 1500 m (4900 ft). This tree is well adapted for growth on rocky outcrops, but it does not grow in the Grand Canyon today. Pinedale climates must have been more to its liking (i.e., cooler and moister than those today), but it may also be excluded today because of competition from ponderosa pine. By about 12,000 yr B.P., limber pine needles drop out of the Grand Canyon midden record, signaling an end of Pinedale climatic conditions. Limber pine appears to have been more abundant throughout the Rocky Mountain region during Pinedale times; it is more marginal there today. Here is a good example of a species that is more successful (i.e., more dominant on the landscape) during glacial intervals than in nonglacial intervals. That might seem like a disadvantage, but remember that perhaps as much as 90% of the last 1.7 million years have been glacial periods. In the Quaternary Period, the chance of survival for a species is probably improved if it has been adapted to glacial environments rather than to nonglacial environments. Thus southwestern plants and animals adapted to cool, moist conditions are just hanging on in microhabitats during the Holocene, waiting their chance to dominate again once the next glacial period comes.

About 3000 years after limber pine disappeared, ponderosa pine came into the Grand Canyon. Its absence from late Pinedale–age middens suggests that regional climates were not suitable for its growth there. In particular, Ken Cole argues that dry summer conditions may have kept it out. Based on its modern distribution in the Southwest, ponderosa pine can tolerate cool summers but not dry ones. Ponderosa pine was more abundant in the Grand Canyon in the early Holocene than it is today. During that interval it extended down the canyon's steep slopes to regions occupied today by piñon-juniper woodland. Perhaps during that time there was an abundance of summer precipitation. Geologists also believe that the last glaciation saw the development of deeper soils in the canyon. Ponderosa pine may have benefited from a combination of these two environmental conditions in the early Holocene. Since then, summers have become drier and erosion has thinned the Pinedale soils.

The comings and goings of these conifer species serve as a reminder that no biological community is truly stable. What we see today is just the latest mixture of species in a given region. Some communities appear to be perfectly adapted to their physical environment, interacting harmoniously with each other. This impression is an artifact of our own creation, founded upon an extremely short period of modern observation compared with the lifespans of the ecosystems.

Modern biologists may be hoodwinked into thinking that biological communities are stable associations of plants and animals because of the short lifespan of humans. Imagine a batch of fruit flies born in June that die in July: they know only the warmth of summer. Ecologically speaking, we humans are like those fruit flies: we know only the warmth of the Holocene and the ecosystems fostered by that warmth; we do not know what the ecological "winter" (i.e., a glacial interval) is like.

Those who know only modern biology might say that I am confusing the biological time scale with the geological time scale. My argument is that those two time scales are nearly the same, as far as ecological events are concerned. The most meaningful scale from which to view ecosystems ranges from centuries to millennia. Changes on the scale of years to decades amount to little more than "noise" in the ecological signal. Since we cannot observe long time scales during our lifetimes as scientists (and funding agencies have never been known to fund an ecological research project than spans several centuries), we are forced to turn to the fossil record to make sense of the current crop of ecosystems. However, the two types of studies (paleoecological and modern ecological) are certainly not mutually exclusive. On the contrary, paleoecologists need the information gathered on modern ecosystem patterns and processes to help them unravel the meaning of the fossil record. And without the fossil record, modern ecologists will never understand the history of modern ecosystems. The radiocarbon time scale (the last 40,000 years) links ecological time and evolutionary time.

Some remarkable records of late Pleistocene vegetation have been preserved in deep lake basins to the south of the Grand Canyon. Scott Anderson studied a 35,000-year pollen record from sediments that accumulated in Potato Lake in Coconino County, about 200 km (125 miles) southeast of the Grand Canyon. Richard Hevley studied a 50,000-year pollen record from Walker Lake, about 100 km (60 miles) southeast of the canyon. Both of these records are amazingly long for a region where few lakes have contained water for tens of thousands of years. Walker Lake lies in the crater of a small cinder cone (ancient volcano). Potato Lake fills a depression in a sandstone formation called a "solution depression," formed as the soft sandstone was dissolved by chemical weathering. This lake yielded almost 10 m of sediments, and Walker Lake yielded 9 m. At Potato Lake, mid-Wisconsin environments (from 35,000 to 21,000 yr B.P.) fostered the growth of an Engelmann spruce forest, mixed with other conifers. Engelmann spruce grows today at elevations at least 460 m higher than Potato Lake, and this situation suggests that summer temperatures were about 5°C cooler then than they are today. During late glacial times, from 21,000 to 10,400 yr B.P., even cooler temperatures prevailed, and Engelmann spruce completely dominated the landscape. Soon after this time, however, climates warmed rapidly, and spruce forest was replaced by ponderosa pine, the dominant tree of the region today. Potato Lake almost dried out during the mid-Holocene, but lake levels increased during the last 3000 years.

The pollen record from Walker Lake takes the regional vegetation history back an additional 15,000 years. From about 50,000 to 45,000 years B.P., cool, mesic conditions prevailed at Walker Lake. This period was followed by an interstadial interval between 45,000 and 35,000 yr B.P. in which climates became substantially warmer and drier. The regional vegetation became quite similar to that seen in the Holocene. Interstadial intervals are times of warming that occur during glaciations. They are not warm enough or long enough to cause continental ice sheets to melt away (i.e., they are not as warm or as long as bona fide interglacial periods), but in this case they caused some melting back of the edges of ice sheets in the north, and they brought warmer, drier climates to the Southwest, if only for a few thousand years.

This particular interstadial warming has been noted in fossil records as far away as western Europe and arctic Alaska, where it goes by various names. Radiocarbon chronologies of this interstadial link the warming event to this particular interval, although the exact timing and duration of the interstadial varied from region to region. Closer to the continental ice sheets, this interstadial ushered in a shift from arctic tundra vegetation to boreal forest, then back again as the warming trend disappeared.

At about 6000 yr B.P., Walker Lake apparently dried up. This pattern matches what was found at Potato Lake. Taken together, the two lake records indicate a mid-Holocene drought of substantial proportions.

The lake sediment pollen records confirm what has been found in packrat midden records from the Colorado Plateau and extend the regional paleoenvironmental record back to 50,000 yr B.P. In a similar vein, pollen records from sediments that accumulated in shallow caves in the Grand Canyon tend to confirm what has been reconstructed from packrat midden records in the canyon.

It is reassuring to paleoecologists to see that different types of proxy data records tell the same story for a given region over long periods of time. The different data sets fill in gaps in each other's records; they add details or help clarify the meaning of certain changes. This is what makes interdisciplinary, regional syntheses exciting to work on.

Late Pleistocene Fauna of the Grand Canyon

The large-mammal fauna that lived in and around the Grand Canyon during the last glaciation was essentially the same as that of the Canyonlands region. Columbian mammoth remains have been found a few kilometers east of the canyon, in one case (a tooth plate) in a condor's nest. Mastodon remains have been found at a site by the Little Colorado River, so it is likely that these proboscideans roamed at least to the rims of the canyon, even if they did not live in the canyon itself. On the other hand, Harrington's mountain goat was well adapted to scaling the high cliffs. Its remains, including dung, horn sheaths, and bones (some with dried tissues still attached) have been found at several localities in the canyon. According to geologist James Mead and others, in one cave impressions where these mountain goats laid down have been preserved for more than 10,000 years! In the same cave, abundant fecal pellets from this species were found. Along the Colorado River, Harrington's mountain goat fed on at least 24 species of grasses, sedges, and rushes. It supplemented its diet with a wide variety of other plants, including the leaves of many bushes and needles of limber pine, Douglas-fir, and white fir. Radiocarbon dates from horn sheaths and fecal pellets indicate that Harrington's mountain goat became extinct shortly after 11,200 yr B.P.

Shasta ground sloth remains have been found in caves in the western Grand Canyon. In Rampart Cave, large amounts of ground sloth dung were found. In fact, until 1976, Rampart Cave contained the thickest and least disturbed deposit of Shasta ground sloth dung known to science. Radiocarbon dates from these dung deposits range from more than 40,000 to 11,000 yr B.P. The tragic loss of information contained in the deposits of Rampart Cave occurred when the dung bed was accidentally set on fire, destroying a fossil resource unique in the annals of paleobiology. Some 200 bones of this sloth were also found in the cave, along with the bones of an extinct Pleistocene horse. Pleistocene packrat middens from this site fill in the botanical history of the immediate vicinity of the cave.

The following reconstruction of the Shasta ground sloth's life-style and its environments in the Grand Canyon has been offered by paleontologist Richard Hansen. Rampart Cave is perched on a steep slope in the canyon's inner gorge. The topographic diversity of the region provided a variety of habitats supporting a rich flora. The slope angles, soil types, and soil moisture levels vary greatly over short distances, creating many microhabitats. Plant macrofossils from the packrat middens show that the landscape surrounding Rampart Cave was dominated by juniper in the late Pleistocene, so the climate was considerably cooler and moister than it is today. Shasta ground sloths fed on a wide variety of plants. Indeed no modern grazer feeds on a larger variety than did this past master from the Pleistocene. The dung pellets of the sloth contained the remains of 72 genera of plants, probably representing at least 100 species. The Shasta ground sloth was undoubtedly a desert gourmand, sampling nearly every type of plant in its range. Most important in its diet (based on the percent composition of plants in the dung) were desert globemallow (52%), Nevada mormon tea (18%), saltbush (7%), catclaw acacia (6%), common reed (5%), various cacti (3%), and yucca (2%). Based on the parts of the plants eaten by the ground sloths, it is apparent that they fed on globemallow, mormon tea, and saltbush throughout the year. They fed mostly on the seed pods and flowers of yucca and agave, and these parts of the plants are available only in the spring. Cactus fruits were eaten in late summer. Interestingly, plants deemed unpalatable by modern-day grazers on the Colorado Plateau (mule deer, mountain sheep, range cattle, and burros) were also unimportant items in the sloth's diet. These relatively noxious plants include creosote bush, snakeweed, juniper, beargrass, and white brittle bush.

There was no significant change in the diet of the Shasta ground sloth just before its extinction at 11,000 yr B.P., and many of the plants in its diet still grow in the Grand Canyon, including six of the most abundant types of plant foods found in the fossil dung. This finding suggests that environmental change at the end of the Pleistocene was not the cause of the ground sloth's demise. Based on its diet and the range of climatic conditions of fossil sites (from hot, dry deserts to cool, moist mountain localities), it would seem that the Shasta ground sloth would manage quite well in the Grand Canyon today. What a shame there are no specimens left to reintroduce this species into the canyon! It would certainly enliven a float trip down the Colorado River if the participants got to see a 500-lb sloth feeding on cactus fruits near the water's edge.

Yesterday's camel was apparently very successful in the Grand Canyon during the last glaciation. Fossil localities have been recorded from both the eastern and western ends of the canyon (Fig. 5.13). Dung deposits from this animal have been found in southern Utah, but not in the Grand Canyon. It became extinct by 11,000 yr B.P. Although bones of both camel and horse have been found in canyon caves, it is not certain that the camels actually lived in the Grand Canyon. The bones may have been carried into the caves by condors.

Pleistocene horses also fared well in the Grand Canyon during the last glaciation. Abundant remains of horses have been found there, as well as in the drainage of the Little Colorado River to the east. The species that dwelled in the Grand Canyon was a burro-sized animal and a burro-sized hoof was found in Rampart Cave. Its living counterpart, the modern burro, shows us how well this type of animal can cope with the narrow ledges and steep paths of the canyon. Not every large mammal inhabitant of the canyon became extinct at the end of the Pleistocene, however. The remains of bison, pronghorn antelope, bighorn sheep, and mule deer are scarce in the Pleistocene fossil record, but they were present, even if in smaller numbers than they enjoy today.

Some modern predators also roamed the canyon region, including gray wolf and mountain lion. Wolves still survive as a species, but they have not been seen around the Grand Canyon for many years. Throughout the American Southwest, as elsewhere, coyotes have expanded their range and numbers as wolves have declined. Wolves have been more systematically hunted, trapped, and poisoned throughout North America. In areas where wolves and coyotes are both found, wolves outcompete coyotes and may even kill their smaller cousins. However, wolves could not survive the eradication efforts of humans, so the wolves' loss has been the coyotes' gain. In this century, human interference has promoted coyotes to the role of top canine predator in many North American ecosystems.

We have taken a look at the hunters and the hunted among the megafauna that lived in the Grand Canyon region during the last ice age, but what of the scavengers? Just as the Serengeti plain of Africa has vultures that clean up the carrion from the landscape, the Colorado Plateau had California condors in the Pleistocene. They scavenged the carcasses of Pleistocene megafaunal mammals throughout western North America. This was their ecological niche on prehistoric landscapes. Paleontologist Steve Emslie studied the remains of condors from eight caves in the Grand Canyon. Most of these caves yielded only a few condor bones, but one, called Sandblast cave, contained partial skeletons of at least five condors, bits of condor eggshell and feathers, and the bones of several megafaunal mammals. The mammal bones were from Pleistocene horse, bison, mammoth, camel, and Harrington's mountain goat. The structure of the condor bones indicates that they came from chicks near fledgling age. So California condors used this cave as a nesting site, and the megafaunal mammal bones probably represent food bones brought to the chicks in the nest. Only small bones of these animals were carried to the cave. These include bones found in the feet, teeth, pieces of leg bone, and vertebrae. These are essentially the same bones brought back to the nest by living condors in California.

The intriguing thing about the condor remains in the Grand Canyon caves is that they are all older than 10,700 yr B.P. In other words, just as the megafaunal mammals became extinct, the condors died out in this region. Radiocarbon dates on condor remains throughout the Southwest show that this big scavenger did not survive after the extinction of the large mammals on which it fed. The California condor only survived in the Holocene on the Pacific coast of California and Oregon. Perhaps coastal populations endured by feeding on carrion of marine mammals and fish that washed up on the shore. This, at least, was a reliable food source even after the extinction of so many of the large land mammals. Twentieth-century human activities, such as the introduction of pesticides that weakened condor eggshells and the reduction of condor habitat through development, may bring these magnificent "thunderbirds" to an end, but it is possible that they were already on their way out because of the elimination of the large mammals whose carcasses they scavenged.

Additional questions remain about this bird. Why did it not flourish on the Great Plains, where huge herds of bison provided a seemingly endless supply of carrion? Why did it not succeed on the coast of the Gulf of Mexico from Texas to

Florida? More research on condor physiology and habitat requirements needs to be carried out, but the numbers of specimens living in the wild are so few that this may never happen. Furthermore, the fossil record indicates that it would be useless to try to reintroduce condors throughout the Southwest; the main food resource that allowed them to survive there in the Pleistocene, carrion of megafaunal mammals, has long since disappeared, so the birds would most likely starve. On the other hand, if cattle ranchers were to leave dead horses and cattle out on the landscape, condors might manage to survive in this region. After all, this is just a variation on the backyard bird feeder, one in which large animal carrion takes the place of sunflower seeds.

Early Peoples of the Grand Canyon Region

The timing of the extinction of such megafaunal mammals as Harrington's mountain goat and the Shasta ground sloth coincides almost exactly with the arrival of Clovis people in the region. Clovis sites have not been found in the Grand Canyon, but there are several nearby (Fig. 6.6). Before the end of the last glaciation, Clovis culture gave way to Folsom.

Folsom people were primarily bison hunters. Though no Folsom artifacts have been found in the Grand Canyon, they have been found in many other localities across the Colorado Plateau, and they are especially abundant near the confluence of the Green and Colorado rivers, in Canyonlands National Park (Fig. 6.6). The fact that human hunters arrived just as so many large mammals became extinct prompts us to ask, as did the poet, "Who killed Cock Robin?"

Ecologist Paul Martin is convinced that these Paleoindian hunters exterminated the Pleistocene megafauna in the New World, a "Pleistocene overkill." The theory proposes that the megafauna of North America was especially vulnerable to Paleoindian hunters because the people were newcomers on this continent at the end of the Pleistocene, and the animals, unaccustomed to human hunters, had little natural fear of them. The Paleoindians that arrived in Alaska via the Bering Land Bridge spread southward as an ice-free corridor opened up between the **Laurentide** and **Cordilleran ice sheets** (Fig. 6.7). Human populations expanded as new territory became colonized, and the megafaunal mammal hunters exterminated many of their prey animals within a few hundred years of their arrival south of the continental ice sheets. Martin's theory holds that the new hunting pressure, combined with rapid climate change, wiped out most of the megafauna on this continent.

Paleontologists, paleoecologists, and archaeologists are sharply divided on the issue of megafaunal extinction. Some believe that humans had little or nothing to do with the extinctions at the end of the Pleistocene. They hold that environmental

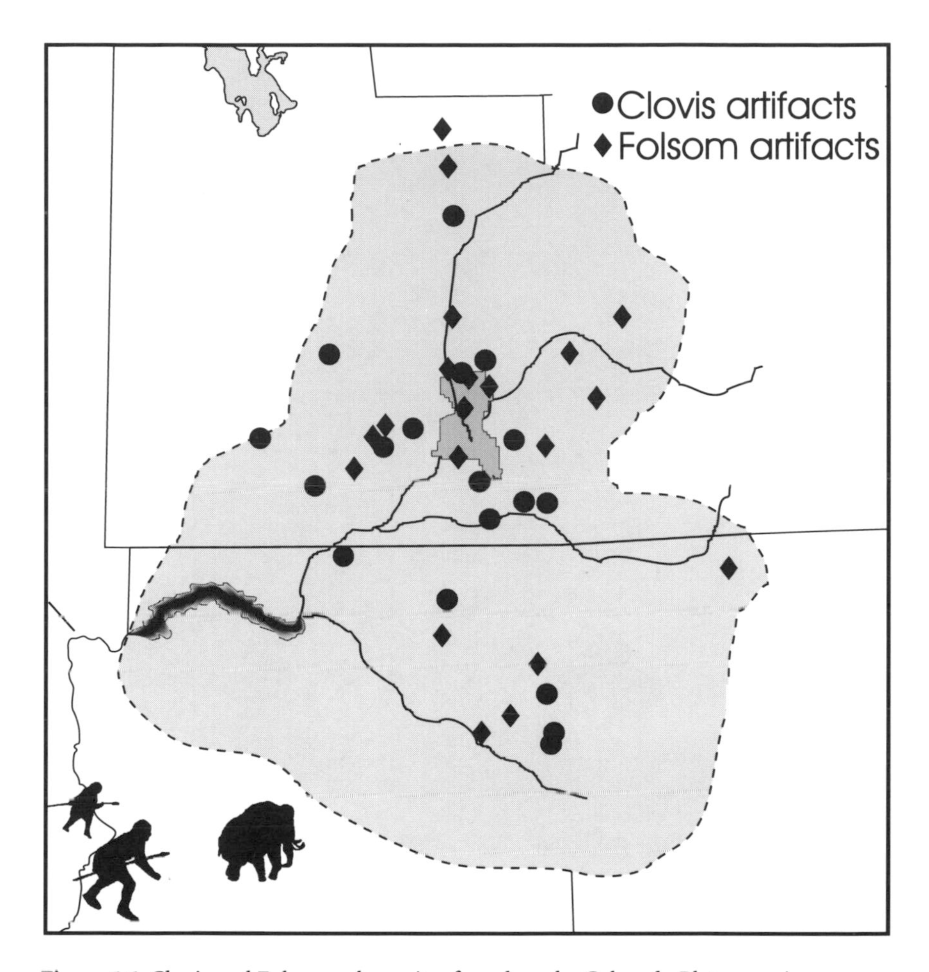

Figure 6.6. Clovis and Folsom culture sites found on the Colorado Plateau region.

change was the catalyst that quickened the demise of the large mammals. Those that support the overkill theory counter this argument with evidence showing that nearly all of the megafauna had survived all the previous glacial-interglacial transitions; that no plants, aquatic organisms, or small animals other than scavengers became extinct; and that the only difference between those climatic change episodes and the last one is the presence of human hunters.

Would you allow me a brief foray into the field of the philosophy of science? Like it or not, our philosophies shape our thinking, and our approach to problem solving. This is as true of scientists as it is of everyone else. One of the fundamental issues at stake in this debate is catastrophism, the idea that the Earth and its biota have been shaped by catastrophic events in the past. This concept developed as a

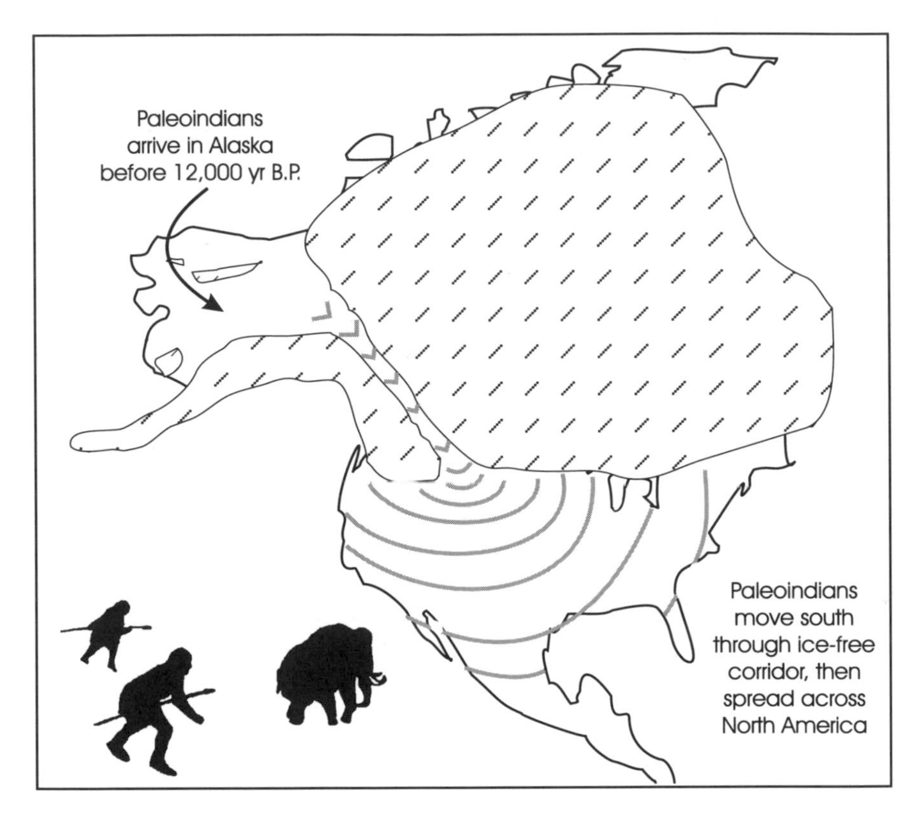

Figure 6.7. Map showing hypothesized rapid spread of Paleoindians from Eastern Beringia to the southern ice-free regions, via an ice-free corridor, starting about 11,500 yr B.P.

scientific argument in the eighteenth and nineteenth centuries. It was linked, for better or worse, with theology. For instance, a catastrophic explanation of the forming of the Grand Canyon would be that Noah's flood, described in the book of Genesis, carved the canyon in a matter of weeks or months. Nineteenth-century scientists rejected catastrophism as fuzzy thinking, not based on observed data. *Webster's New World Dictionary* defines science as "systematized knowledge derived from observation, study, and experimentation," so scientists have tended to reject all catastrophic theories explaining past events because of catastrophism's historical link with theology.

Unfortunately, they may have thrown the baby out with the bathwater. Catastrophes happen. They have happened in the past, just as surely as they will continue to happen in the future. For instance, since our planet is orbiting the sun, which is also orbited by asteroids and comets, sooner or later Earth will intersect the path of an asteroid or comet large enough to do serious damage to our planet.

One only has to look at the moon to see the effects of thousands of such impacts. Earth's atmosphere, chemical weathering, the action of water, and soil erosion tend to blur the impact craters over time, but to deny that such collisions occur is to deny Newtonian physics. Still, many scientists remain adamantly opposed to the theory proposed by Louis Alvarez that dinosaurs were brought to extinction by an asteroid impact on the Earth, in spite of the physical evidence that supports this theory. It may be the theory that best explains all the available facts about dinosaur extinction, but it smacks of catastrophism, so some scientists reject it out of hand. Likewise with Paul Martin's theory about megafaunal overkill.

We will probably never know if the overkill theory is the right one, since the fossil evidence is so spotty. However, if it does the best job of explaining the data we have accumulated, it should at least be given full consideration. Besides the rejection of catastrophism, there are other subtle forces at work here as well. Some archaeologists reject the overkill theory because it challenges their image of Paleoindians as the New World's first ecologists, living in perfect harmony with nature. Such a noble race would never kill more animals than absolutely necessary. This concept of the Paleoindians does not consider that human beings everywhere and at all times in the past have responded to their environments in much the same way. Given adequate tools and sufficient numbers, people have always exploited natural resources (including game animals, minerals, water supplies, and living space) to the point where some of those resources became extremely scarce. This is the curse of our species: we are intelligent enough to devise many ways of exploiting our environment, but apparently not wise enough to stop before we overexploit our environment. Any excursion through the pages of human history will reinforce this fact; why should we suppose that Paleoindians were any different? Even though scientists like to think of themselves as purely rational, logical thinkers, they cannot really help but bring their own emotional baggage to the discussion table. Martin's theory may or may not be true, but we have at least to consider it. On the other hand, like the asteroid impact theory of dinosaur extinction, Martin's Pleistocene overkill theory has been controversial enough to spawn a lively debate and send researchers scuttling off in many directions to find more facts that either refute or support the theory. Perhaps this is the finest compliment that can be paid to a scientific theory—that it motivated dozens of people to go out in search of more evidence, pro or con, thus stimulating research in a given field for decades to come.

Whatever the cause, we are left with a rather poor collection of large animals that have made it through the Holocene. Only one-third of the species that were on the land in the Pleistocene are now living in North America. Nevertheless, they are true survivors. If they were human, they would probably sport hats and T-shirts with the motto "We survived the Pleistocene-Holocene transition!"

Before moving on to postglacial times, let me pose an additional question. Why do we not have a national park or monument dedicated to the fascinating peoples of the Clovis and Folsom cultures? Both of these cultures were discovered in New Mexico, and many of their sites have since been unearthed in the American Southwest. Why has the government overlooked these groups? Surely we do not still cling to a Eurocentric view that Columbus discovered America, and that the indigenous peoples were unimportant? In the late twentieth century, we seem to have gotten beyond that outmoded belief, but what of the very first peoples to enter North America? Think of the hardships and dangers they faced! Perhaps it is still not too late for the creation of a "Clovis and Folsom National Park." It might have to be a series of properties managed by the National Park Service, including Blackwater Draw, New Mexico (the site of the original Clovis discovery), Folsom, New Mexico, and a few other key sites such as Dent, Colorado, and Domebo, Oklahoma, where Clovis campsites and the remains of butchered megafaunal mammals have been found. Amazingly, the Blackwater Draw locality has been vandalized and looted through the past 50 years. Teenage boys have used the slopes where mammoth bones are still buried as racetracks for motorcycles. Although this abuse of one of the most important archaeological sites in the world has been curtailed in recent years, this site, and many others, have nevertheless not been given the protection and acclamation due to them. The creation of a national park or monument at these sites would go a long way toward redressing this oversight. It would also educate the public, making them aware that some fascinating peoples lived in this region long before the Anasazi, the only group of prehistoric peoples currently honored with their own national parks and monuments in the American Southwest.

Postglacial Transition: Changes in Regional Ecosystems

In contrast to the Pleistocene megafauna, the vegetation of the Grand Canyon region did not lose species by extinction at the end of the Pleistocene. Rather, species moved around and became settled in new regions and elevations. Imagine a troupe of plant "players" on the Pleistocene stage of the Southwest. The number and kind of players seem to have been fixed long ago, perhaps half a million years ago or more. No new players have arrived since then, and none have left. During any given "act" of the late Quaternary "play," only certain members of the cast occupy center stage. These players dominate the scene for a while, then are replaced by others who have been waiting in the wings, sometimes for thousands of years! As in the human theater, today's star will be the has-been of tomorrow. Today's chorus-line member will be tomorrow's showstopper. The Pleistocene vegetation drama had its last curtain call about 11,000 yr B.P., but by then some of the previously important players had already departed the scene. Since more than 90% of the last 1.7 million years has been spent in glacial intervals, the cool-moist–adapted species have enjoyed more success overall than their warm-dry counterparts. The current interglacial warmth is but a brief respite from the dominant theme of continental ice sheets to the north, cool-moist climates to the south. Again, because of our short human lifespans, we tend to look upon the current conditions as "normal." In fact, the ecosystems of the world have gone through at least 17 glacial-interglacial cycles during the Quaternary. Interglacial periods tend to last anywhere from 10,000 to 15,000 years, whereas glacial periods can last more than 100,000 years. How do plants cope with these massive climatic upheavals? Do they keep pace with environmental change, or do they lag behind? Evidence from the Grand Canyon has been particularly useful in answering these questions.

In his study of packrat midden plant records from the Grand Canyon, Ken Cole came to the following conclusions. First, the late Pinedale vegetation of the canyon was reasonably stable. Until about 12,000 yr B.P., few species migrated into or out of the region. After 12,000 yr B.P., however, major disruptions occurred. The transition between Pinedale and Holocene climates brought about the most radical changes in regional vegetation recorded in the packrat midden record. Some species of plants that had done well during the Pleistocene were eliminated from the canyon. Ken Cole's data show that there is a noticeable gap between the time of departure of some of these species and the time of arrival of their Holocene "replacements" (Fig. 6.8). Early Holocene plant communities were not like their predecessors, nor were they like those that came after. As I mentioned briefly earlier in this chapter, the early Holocene appears to have been a time of increased summer rainfall. This trend gave a competitive advantage to the species that grow well under these conditions, such as ponderosa pine. However, Cole believes that the unique mixture of plant species that occupied the canyon during the early Holocene was due more to plant species being out of equilibrium with the rapidly changing climate. This disequilibrium, or disharmony, was brought about because many species of plants cannot migrate out of old regions and into new ones fast enough to keep up with climate changes, especially when the climate changes as rapidly and drastically as it did during the Pleistocene-Holocene transition. If you want to find out how fast and how much the climate changed in that tumultuous period between 13,000 and 10,000 yr B.P., you must look at other lines of evidence.

During the past 15 years, I have been developing a reconstruction of climate change for the Pleistocene-Holocene transition based on fossil insect data from the western United States. This story is explained more fully in my book, *The Ice-Age History of National Parks in the Rocky Mountains.* However, as mentioned previously, at 14,000 yr B.P., regional summer temperatures were about 10°C (18°F)

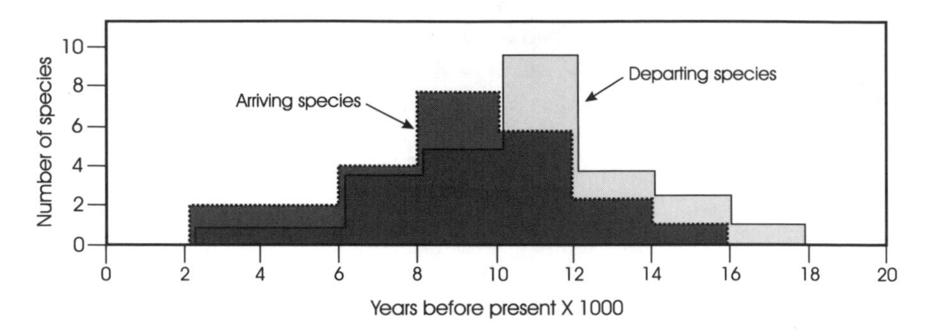

Figure 6.8. Arrivals and departures of plant species from the Grand Canyon packrat midden fossil record during the last 20,000 years. (After Cole, 1990.)

cooler than they are today. Between then and 11,000 yr B.P., summer temperatures in the West rose about 5°C (9°F). Between 11,000 and 10,000 yr B.P., summer temperatures rose an additional 5°C. In other words, about half of the climatic warming that marked the transition from glacial to interglacial climates took place in just 1000 years. Paleobotanical evidence from the Grand Canyon suggests that Pinedale summers there were cooler by only about 6–7°C (11–13°F) than they are today. Precipitation may have played a greater role in controlling biological communities. Using shifts in vegetation zones between past and present elevations, Cole estimated that the north rim received as much as 41% more moisture per year during late Pinedale times and the south rim received as much as 24% more than it does today. Of course, differences in atmospheric circulation patterns during the Pinedale probably caused differences in the seasonality of the precipitation (i.e., most of the precipitation may have fallen as snow in winter, or it may have fallen as rain in the spring, summer, and fall).

Many species of plants, especially long-lived plants such as conifers, responded quite slowly to these climate changes. In some cases, species that no longer fit the regional climate hung on for up to 1000 years after the climate became unsuitable. Some conifers have the ability to switch off their sexual reproduction (i.e., they stop producing pollen with viable sperm cells or cones with viable seeds). They may survive very great lengths of time by vegetative reproduction, sending up shoots that are just new parts of the original individual. In other cases, the Pleistocene inhabitants died out, but no new species migrated in to take their place until hundreds or even thousands of years had elapsed. Other species did migrate rapidly, and the combination of species at a given site was a mixture of rapidly and slowly migrating plants. Not too surprisingly, some of these mixtures do not make ecological sense when viewed as a community. In other words, our knowledge of their ecological requirements would suggest that these species do not belong together, since they require different and even mutually exclusive conditions for establishment, growth, and reproduction.

Some paleobotanists have argued that it was the unique character of late Pinedale and early Holocene environments that led to the temporary creation of these misfit plant communities. Their arguments are not without merit, but they also fail to solve all of the problems Here is one of the unsolved problems. When climates became warmer at the end of the ice age, cool-adapted plant species had to migrate either north or upslope to find suitable conditions. In the Grand Canyon, plant species migrated very slowly northward following the late Pinedale, whereas they were able to migrate rapidly up the walls of the canyon. Obviously, the trip up the canyon was far easier to accomplish than a trip to new latitudes. But if all plants respond rapidly to climate change and are able to remain in equilibrium with those changes, then we would expect to see no significant differences between the time it took to move upslope in the Grand Canyon and the time it took to migrate into the mountains of central Utah. Cole's data show that distributional shifts in long-lived plant species lag behind climate changes, so that at the end of the Pleistocene, regional plant communities were not in equilibrium with changing climates. These results make many paleobotanists unhappy, but they do not alter the validity of Cole's conclusions, in spite of many harsh criticisms.

Holocene Ecosystems

The Grand Canyon midden record indicates that most warm-adapted species had finished migrating into the canyon by the mid-Holocene. However, some of the most warmth-loving species, such as ocotillo, did not become established there until very recently. Early postglacial plant macrofossil assemblages come from packrat midden samples that date from between 11,000 and 8500 yr B.P. These fossil assemblages are similar in composition to associations of plants now found along the Mogollon Rim, south of the canyon. Plant species associated with modern chaparral communities were found in early Holocene middens from canyon rockshelters situated in Coconino Sandstone and Redwall Limestone formations. The nearest modern community comparable to these early Holocene assemblages can be found near Sedona, Arizona. Early Holocene climates in the Grand Canyon were marked by temperatures similar to those found there today, but with increased summer precipitation.

During the middle Holocene, the paleobotanical record from Grand Canyon packrat middens reflects increasing environmental stresses on plant communities, probably due to a combination of decreasing precipitation and increased evaporation. The lower limits of juniper and piñon pine shifted up the canyon to higher elevations. The environmental changes needed to bring about this type of shift are not very great. An increase of only 1°C (2°F) in average temperatures, combined with slightly less precipitation, is all that would be required. Modern environmental conditions became established by about 1200 yr B.P., and plant communities in the Grand Canyon have been relatively stable since that time.

Archaic and Anasazi Cultures in the Grand Canyon

Archaeology is not the main focus of most visitors to the Grand Canyon. Nevertheless, people have been living in and around the canyon for several thousand years, beginning with the temporary hunting camps of Clovis and Folsom peoples in this part of the Colorado Plateau. During the Archaic Period, people began occupying sites in the canyon on a more frequent basis. One of the most fascinating sets of Archaic artifacts that has been discovered in the Southwest is a series of figures made of split willow twigs, twisted and bent into the shapes of animals (Fig. 6.9). In 1933, members of a Civilian Conservation Corps crew working in the canyon found three of these miniature figures in a cave. Each figure had been made by bending and folding a single willow twig that had been split down the middle. The workers had no idea of the age of these figures; they thought that they had been made by Indians, so the figures were set aside for future study or display in the park. Since then, more than 300 of these objects have been found. They have all come from caves and rockshelters, perched in seemingly inaccessible places along the canyon walls. Most are made of willow, but some are made of cottonwood twigs. Most are thought to represent deer or bighorn sheep; others may represent antelope.

These figures may have been fashioned for use in religious ceremonies or as magical aids in hunting. Before you scoff at this seemingly primitive notion, remember that hunting is, by its very nature, a hit-and-miss venture. Archaic Period hunters relied heavily on game animals for their survival. Hunting was not their hobby; it was a necessity. Human nature has not changed all that much, either. Many hunters, both ancient and modern, have taken steps that they believed would improve their luck. These activities range from preparing figurines to wearing a "lucky" shirt on the opening day of the hunting season. The split twig figurines from the Grand Canyon have been radiocarbon dated. Their ages fall between 4000 and 3000 yr B.P. Whereas most artifacts made of vegetable matter have long since decomposed, these little treasures of antiquity survived in their high, dry caves along the canyon walls.

Recently paleontologists Steve Emslie, Jim Mead, and Larry Coats discovered additional "shrine" sites, perched high up on canyon walls in almost inaccessible

Figure 6.9. Split twig figurine from the Archaic period, found in the Grand Canyon.

sites in the Grand Canyon. The sites had not previously been surveyed for archaeology or paleontology because it took climbing ropes to get to them. Radiocarbon dates indicate cultural use of these caves between 4390 and 3700 yr B.P., during the middle of the Archaic Period. Unlike the caves that had only split-twig figures, these caves had numerous cairns, set out in rows. Some of the cairns were made of only of piles of rock; others were made of combinations of rocks and pieces of late Pleistocene packrat midden that had been dug out of larger middens in the cave. The most remarkable aspect of these recent finds is that the Archaic people who built these shrines also incorporated horn sheaths from the extinct Harrington's mountain goat, along with split-twig figures. The placement of the cairns in the caves followed no set pattern, so it appears that individuals, not groups of people, set up these shrines. The mountain goat remains were probably found in these or other nearby caves in the Grand Canyon. Did the Archaic people who found them

Figure 6.10. Anasazi ruin near the foot of the Grand Canyon. (Photograph courtesy of Corel Corporation.)

know that they represented fossil remains of extinct species? We cannot be sure of this, but the people who used the mountain goat remains in setting up their shrines obviously considered the mountain goat fossils unusual. If the Archaic people realized that the horn sheaths were unlike any they had seen before (and this seems likely, given the level of familiarity these hunter-gatherer people must have had with regional wildlife), then they may have thought that the fossils had magic powers. It sparks the imagination to realize that the first people to appreciate the megafaunal mammal fossils in the Grand Canyon were not twentieth-century paleontologists but Archaic Indians who lived in the mid-Holocene.

The Grand Canyon was a crossroads for a variety of late prehistoric cultures, especially the Anasazi. Other cultures were only marginally represented in the Grand Canyon, but Anasazi village ruins are the most extensive in the park, having been found on both the north and south rims, as well as down near the river along canyon walls (Figs. 6.10 and 6.11).

The first evidence of permanent farming settlements in the Grand Canyon region comes from the south rim, where farmers began working about 700 A.D. By 900 A.D., land on the north rim was being worked. Over the next 150 years, agricultural activities expanded, and the number of communities grew accordingly. A good example of a small farming community on the south rim is preserved at the Tusayan ruin, a few kilometers east of the visitor center along the road to the east entrance.

Pueblo people began living in the canyon about 1050 A.D., growing crops on nearly every patch of arable land in the canyon. However, this agricultural experiment was doomed to fail because of a lack of moisture. The Grand Canyon is a region that is marginal for agriculture, even in the best of times. Even years of average precipitation may not have been sufficiently wet to allow prehistoric peoples to farm there.

By 1150 A.D., not only had settlements in the canyon been abandoned, but agriculture on the north and south rims had also failed. The only settlement to survive the drought that doomed these sites was that in and around Havasu Creek, one of the most consistent sources of water near arable land in the region. Agriculture and human occupation have continued there to the present time, but never again would people try to farm other regions of the Grand Canyon.

When looking into the Grand Canyon from its rims, one is tempted to ask, "How could anybody farm down there?" Even archaeologists were skeptical—until they began digging up ancient settlements along the canyon floor. Although the inhabitants of those communities were relatively isolated from the world above the canyon walls, they did manage to maintain substantial contact with other peoples.

Figure 6.11. Nankoweap Canyon, Grand Canyon National Park. (Photograph courtesy of Corel Corporation.)

A system of trade brought pottery from a variety of other regions into the Grand Canyon. We are left with a sense of appreciation for their tenacity at making a living in that seemingly inhospitable region.

I hope that this chapter will inspire the reader to take a closer look at the Grand Canyon. Upon first inspection from the rim, it looks like a huge, sterile slice through bedrock. Life in the canyon appears to be limited to the banks of the Colorado River, far below. You have to look a little harder to find the plant and animal communities of the canyon, but your search will be rewarded as you gain an appreciation for the persistence of life in this rocky domain. The biological communities of the canyon were perhaps more conspicuous during the last ice age. The forests that clothe the rims today crept down the canyon to great depths, and large mammals roamed both the rims and the rocky ledges, leaving extraordinary evidence of their existence in dry caves in the canyon walls. The geology of the canyon will always get top billing, because the canyon itself is one of the great natural wonders of the world. But there is another fascinating story to be learned here. Dry caves of the Grand Canyon have yielded some of the most remarkable perishable remains of ice-age fauna and flora. These wonderful fossils tell the story of life during the last 40,000 years, played out with the canyon as a backdrop. In this story, the trumpeting of a mammoth herd echoes back from canyon walls, and a lone hunter scales a tall cliff to place a small figure of bent twigs in a hidden cave, perhaps an offering to his gods.

Suggested Reading

Anderson, R. S. 1993. A 35,000 year vegetation and climate history from Potato Lake, Mogollon Rim, Arizona. *Quaternary Research* 40:351–359.

Cole, K. L. 1990. Late Quaternary vegetation gradients through the Grand Canyon. In Betancourt, J. L., Van Devender, T. R., and Martin, P. S. (eds.), *Packrat Middens: The Last 40,000 Years of Biotic Change.* Tucson: University of Arizona Press, pp. 240–258.

Emslie, S. D. 1987. Age and diet of fossil California condors in Grand Canyon, Arizona. *Science* 237:768–770.

Emslie, S. D., Mead, J. I., and Coats, L. 1996. Split-twig figurines in Grand Canyon, Arizona: New discoveries and interpretations. *Kiva* 61:145–173.

Euler, R. C. (ed.). 1984. *The Archaeology, Geology, and Paleobiology of Stanton's Cave, Grand Canyon National Park, Arizona.* Grand Canyon Natural History Association Monograph 6. 141 pp.

Hansen, R. M. 1978. Shasta ground sloth food habits, Rampart Cave, Arizona. *Paleobiology* 4:302–319.

Harris, A. G. 1977. Grand Canyon National Park. In *Geology of the National Parks.* Dubuque, Iowa: Kendall/Hunt, pp. 3–14.

Hevly, R. H. 1985. A 50,000 year record of Quaternary environments, Walker Lake, Coconino County, Arizona. *American Association of Stratigraphic Palynologists Contribution Series* 16:141–154.

Lister, R. H., and Lister, F. C. 1983: *Those Who Came Before.* Globe, Arizona: Southwest Parks and Monuments Association. 184 pp.

Martin, P. S., and Klein, R. G. (eds.). 1989. *Quaternary Extinctions.* Tucson: University of Arizona Press. 892 pp.

Mead, J. I., O'Rourke, M. K., and Foppe, T. M. 1986. Dung and diet of the extinct Harrington's mountain goat (*Oreamnos harringtoni*). *Journal of Mammalogy* 67:284–293.

Nelson, L. 1990. *Ice Age Mammals of the Colorado Plateau.* Flagstaff, Arizona: Northern Arizona University Press. 24 pp.

O'Rourke, M. K., and Mead, J. I. 1985. Late Pleistocene and Holocene pollen records from two caves in the Grand Canyon of Arizona, U.S.A. *American Association of Stratigraphic Palynologists Contribution Series,* 16:169–186.

Withers, K., and Mead, J. I. 1993. Late Quaternary vegetation and climate in the Escalante River basin on the central Colorado Plateau. *Great Basin Naturalist* 53:145–161.

7

THE ANASAZI RUINS
Mesa Verde and Chaco Canyon

Mesa Verde: Plateau in the Sky

Mesa Verde is the only American national park devoted to the works of prehistoric peoples. When visitors stand on top of the canyon and gaze down into the opposite side to see Cliff Palace, their reactions often match those of author Willa Cather, who wrote, "I saw a little city of stone asleep. . . . That village sat looking down into the canyon with the calmness of eternity . . . preserved in the dry air and almost perpetual sunlight like a fly in amber, guarded by the cliffs and the river and the desert." That is one of the wonders of Mesa Verde: the feeling that you might just be the first person to see these cliff dwellings since they were last inhabited by the people who made them; they look so fresh, so undisturbed, so immune from the ravages of time. The presence of cliff dwellings and other ancient structures at Mesa Verde was known to the local Indian tribes, the Utes and Navajos, for a long time before their discovery by Europeans, but the native peoples who came after the Anasazi left their ancient villages alone. Who can blame the Utes, Navajos, or other natives if they felt spooked at Mesa Verde. Surely nearly every visitor to these ancient habitations feels as if there might still be ghosts of the original inhabitants wandering about, peering through the tiny windows, hiding in a kiva, or traipsing down a watercourse in search of game. I think these feelings well up in us because, in spite of what we might know about the antiquity of the place, the villages at Mesa Verde *look* as if they were just deserted a few days ago.

The unique geographic setting of Mesa Verde is probably what attracted ancient peoples here in the first place. Even more ancient events shaped the landscape, soils, vegetation, and animal life of the region. In a geologic sense, the Anasazi were latecomers on the scene. Let us look at the physical setting and more of the ancient history of the region, in order to place the Anasazi occupation of Mesa Verde in a proper context.

Much has been written about the artifacts and architecture of Mesa Verde. Most people know less about the ancient environments of the mesa and how paleoecological tools have been brought to bear to address some of the questions that pottery sherds and brick walls cannot answer. My main goal in this chapter will be to tell the story that developed by using paleoecological tools: packrat middens, insect remains, human coprolites, and tree ring studies.

Modern Setting

Mesa Verde is a large plateau region in southwestern Colorado, dissected by several steep-walled canyons carved by streams (Figs. 7.1 and 7.2). At Park Point, the highest point in the park, you stand at more than 2600 m (8570 ft) elevation above sea level and more than 600 m (2000 ft) above the Mancos Valley floor to the east. Such high elevations in southern Colorado are generally quite frigid for much of the year, with late-lying snow in spring and early frosts in fall. But Mesa Verde is unusual, and the Anasazi people had good reason to settle here. The climate of Mesa Verde is substantially different from that of the surrounding lowlands and mountains just to the north. Some of the unique aspects of this climate have probably been in place since the end of the Pleistocene. According to climatological data collected in this century, Mesa Verde has warmer summer temperatures than it should based strictly on its elevation. This is because of the peculiarities of cold air drainage and the angle at which the mesa is inclined. The mesa tops slope southward, collecting more sun than adjacent flat regions (Fig. 7.1). The mesa also has less variation between daytime and nighttime temperatures than almost anywhere else in southwestern Colorado. These temperature anomalies add up to a longer growing season for Mesa Verde than for most other localities in the Four Corners region. On top of that, Mesa Verde receives abundant moisture almost year round. During eight or nine months of the year, Mesa Verde receives an average of 400 mm (15 in.) of precipitation. It is especially abundant in late winter and later summer, when monsoon moisture arrives from the Gulf of California.

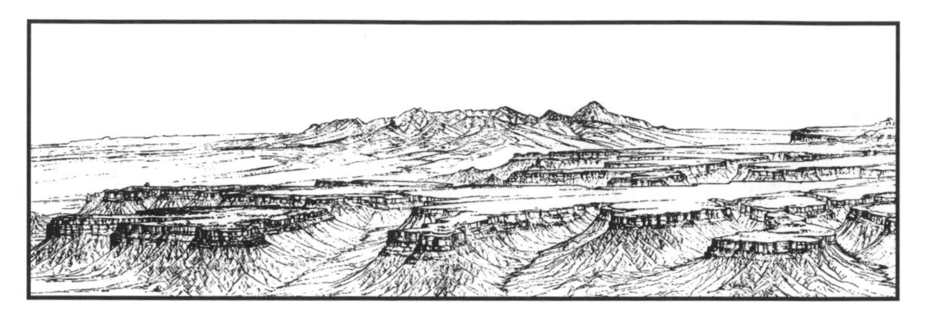

Figure 7.1. Aerial view of Mesa Verde, showing dissection of the plateau by numerous canyons. The view is from the south; note how the mesa tops slope towards this southern direction. (From Powell, 1895.)

These factors combine to make Mesa Verde especially suitable for agriculture. It is a little island of moderate, moist climate, nearly surrounded by semiarid to arid lowlands where sustaining agriculture is difficult today and would have been even more arduous for prehistoric peoples with very limited technology.

The geology of Mesa Verde also contributed to the success of Anasazi habitation. The canyons that dissect the mesa cut down through sandstone, a porous rock that allows water to seep down from the top of the mesa to near the bottoms of the canyons. Beneath the sandstone is a layer of shale that is almost impermeable to water. Where the water meets the shale layer, it emerges out of the rock in the form of seeps and springs, such as the one at the head of the canyon at Spruce Tree House. These seeps and springs are generally reliable, year-round sources of water. As moisture in the sandstone layers freezes in winter, it creates cracks that loosen the sandstone, causing it to break off and fall into the canyon. Over many millions of years, this process brought about the overhangs in the cliff walls where the Anasazi built their cliff dwellings.

About 45% of the Colorado Plateau lies below 1850 m (6070 ft) elevation; much of this landscape is covered by piñon-juniper woodland (Fig. 7.3). The modern vegetation of Mesa Verde National Park is dominated by this woodland. It is the principal vegetation type of the entire Colorado Plateau region, at elevations between 1600 and 2100 m (5250 and 6900 ft). Below 1600 m (5250 ft), it is replaced by desert scrub vegetation, including sagebrush, shadscale, blackbrush, and various grasses.

At the lower elevational boundary of piñon-juniper woodland, the trees are small and widely spaced. With increasing elevation, they become larger and are set closer together, forming a dense woodland near the upper end of their elevational range, where they also mix with ponderosa pine, Douglas-fir, and Gambel oak. In

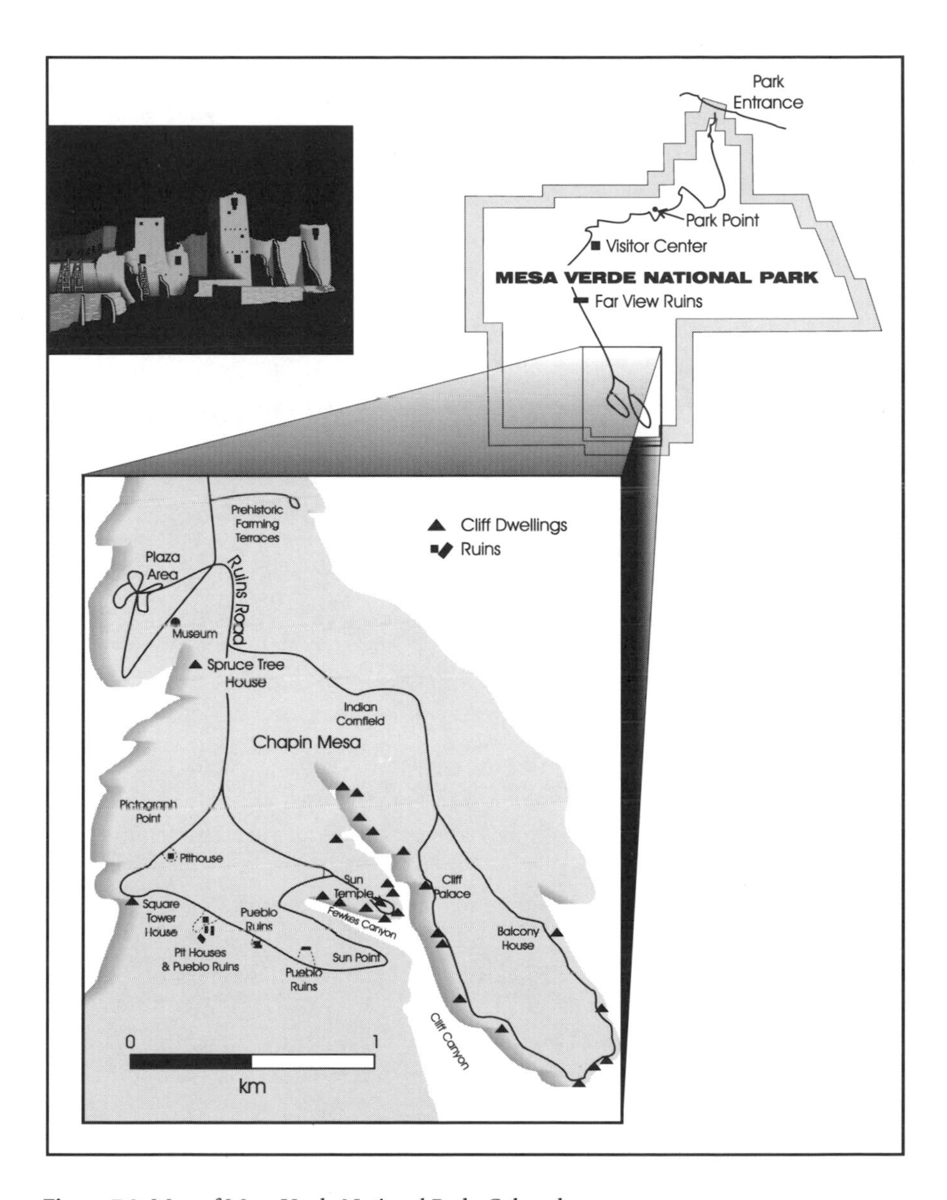

Figure 7.2. Map of Mesa Verde National Park, Colorado.

Mesa Verde as elsewhere, much of the ground under the trees is bare and rocky or supports a sparse cover of shrubs and grasses. Some of the wild plants most important to the Anasazi were the prickly pear cactus and yuccas. Prickly pear fruits and cactus pads were harvested for food. Cactus spines were removed by searing over a fire. The jointed stems of cholla cacti were roasted and stored for

Figure 7.3. Pinyon-juniper woodland, Arches National Park. (Photograph by the author.)

later use. Yucca fibers were woven into baskets, ropes, mats, sandals, and other articles of clothing. The sharp tips of yucca leaves were used as needles for sewing. Yucca flowers were harvested for food, as was the plant's starchy fruit, which resembles a cucumber. Yucca roots yield a soap that was used as a shampoo.

Seeds and fruits of wild plants supplemented the diets of the Anasazi and have been a vital food resource to regional hunter-gatherers throughout prehistory. Some of the most important wild plant foods included wild onion bulbs, wild potato tubers, seeds and seed pods of evening primrose, mesquite, hackberry, goosefoot, lamb's-quarters, Rocky Mountain bee plant, bunchgrass, Indian rice-grass, and sunflower.

Some of the shrubs that grow on Mesa Verde were also important to the Anasazi culture. Utah serviceberry is a large shrub that yields berries that were dried in the fall. Gambel oak produces acorns that were likewise used for food, after prolonged boiling in water to remove some bitter, toxic chemicals. Gambel oak wood is relatively hard and durable, compared with local conifer woods. It was used to make digging sticks, bows, and other tools.

Juniper and piñon pine were also important sources of food and fiber to the Anasazi. Piñon pine cones contain edible seeds, pine nuts, that were harvested from fallen cones and stored for use in winter. They were boiled, roasted, or eaten raw. The gray-green berries (modified cones) of juniper were used to season foods, and shredded juniper bark was used for a number of things, including baby diapers. These two trees were also the main source of firewood for the cliff dwellers. The inner bark of juniper was sometimes chewed as an emergency food.

Along the streams in the canyons, broadleaf trees and shrubs are found. These include willow and cottonwood. More luxuriant patches of shrubs and grasses are also supported by the abundant moisture along these riparian corridors. These well-vegetated places remain the frequent haunts of mule deer and other game animals.

During late glacial times, subalpine woodlands (dominated by subalpine fir and Englemann spruce) and montane forests (with Douglas-fir, Colorado blue spruce, limber pine, and white fir) crept downslope to occupy elevations now clothed in piñon-juniper woodland, and piñon-juniper woodland shifted downslope to many valleys now covered by desert scrub and grasses.

Pre-Anasazi Cultures

The rich Anasazi culture that produced the cliff dwellings at Mesa Verde and the imposing city at Chaco Canyon did not spring full-blown onto the Colorado Plateau. Rather, the classical Anasazi period was the end product of thousands of years of cultural and technological development, starting with the Clovis culture near the end of the last glaciation (Table 5.1). These people were probably the direct descendants of nomadic hunters who entered the New World from Siberia by way of the Bering Land Bridge, sometime before 12,000 yr B.P. Clovis hunters made distinctive projectile points (Fig. 7.4) that archaeologists have used to classify their culture. Clovis people were present in Cochise County, southeastern Arizona, by 11,300 yr B.P. There are several mammoth-kill sites along the San Pedro River that document their hunting activities there. The youngest Clovis artifacts yet found in North America date from about 11,000 yr B.P.

With a documented span of only 500 years, this culture was relatively short lived. One reason is that Clovis peoples appear to have specialized in mammoth hunting. Mammoth bones have been found in nearly every Clovis site throughout the Southwest, along with bones of bison, Pleistocene horse, camel, tapir, wolf, bear, and jackrabbit (the latter would seemingly have been a light snack in a mammoth hunter's diet). Most of the prey animals taken by Clovis peoples became extinct shortly after the time that the traces of Clovis culture disappeared (give or take a few hundred years). This extinction included several kinds of mammoth, North American camel, ground sloth, Pleistocene horse, saber-toothed cat, North American lion, giant short-faced bear, and giant Pleistocene bison. Perhaps this was one of the important forces behind cultural change in Paleoindian bands at the end of the last ice age.

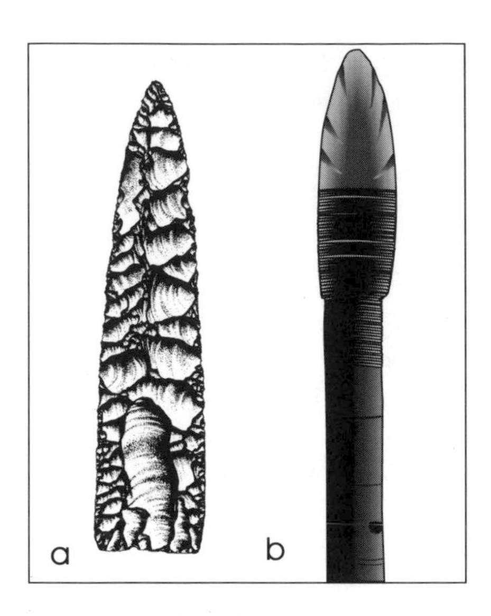

Figure 7.4. (a) Clovis projectile point from the Anzic site, Wyoming. (After Frison, 1991.) (b) Method of hafting (mounting a point onto a wooden shaft).

Clovis hunters apparently specialized in big-game hunting, and some researchers believe that mammoths were their favorite prey animal. Clovis-age artists may have carved petroglyphs depicting mammoths in at least two localities on the Colorado Plateau. Others believe that Clovis hunters did not kill mammoths as often as they simply scavenged mammoth carcasses. Although scavenging seems a less noble occupation, it would certainly have been less dangerous! Attacking a full-grown mammoth with **atlatls** or spears would have been an extremely hazardous endeavor, for a wounded mammoth was surely an angry mammoth.

Many mammoth bones have been found in ponds, lakes, and bogs throughout North America. This fact has lead to additional speculation about mammoth hunting by Clovis peoples. Did the hunters ambush mammoths when they went to their favorite watering holes to drink? A mammoth mired in wet mud would have been easier to subdue than a mammoth running across a dry plain. Or perhaps old, sick mammoths lingered near the water, abandoning the herd as they neared death. Hunters could have dispatched such an animal, or simply waited for it to die, then butchered it near the pond. A third theory, put forward by paleontologist Daniel Fisher of the University of Michigan, is that Clovis hunters killed mastodons and mammoths, then cached large portions of the meat (such as hindquarters, whole legs, and ribs) in ponds during the winter; the pond ice would thus preserve the meat until spring. Fisher's theory makes good sense, and it takes into account the fact that a mastodon or mammoth carcass would yield a huge quantity of meat

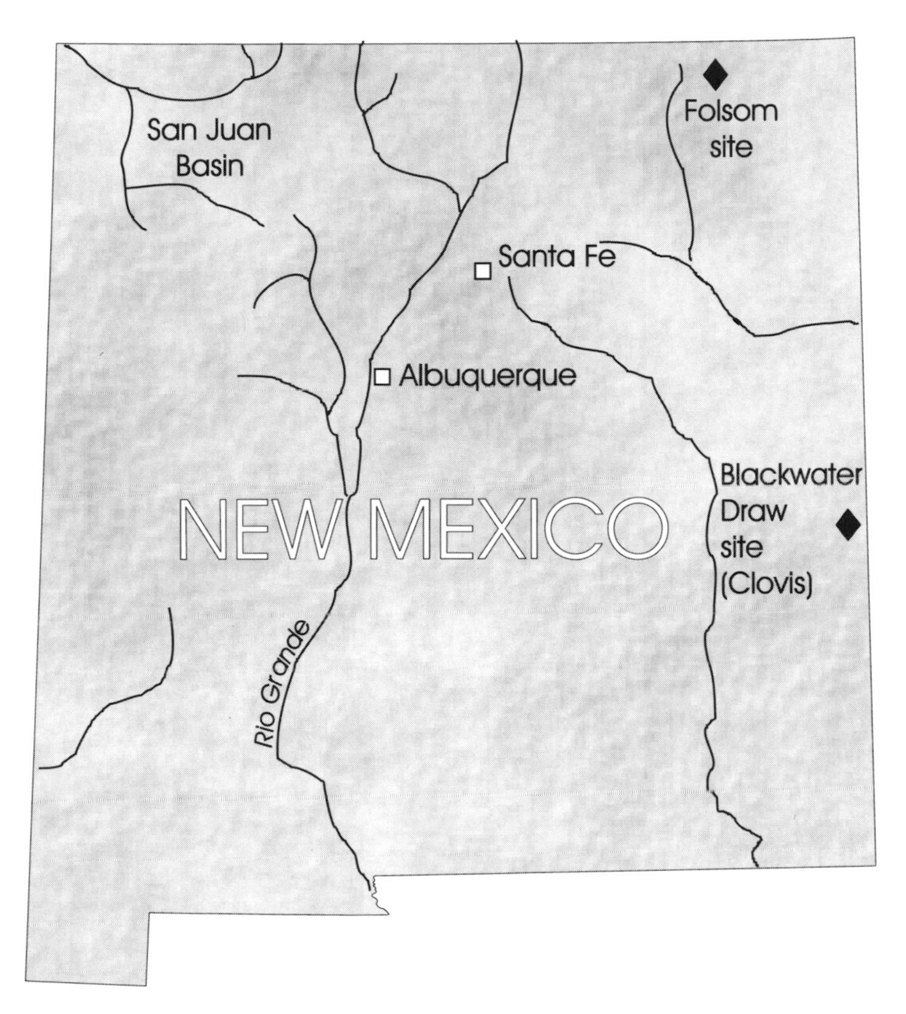

Figure 7.5. Map of New Mexico, showing location of the Clovis (Blackwater Draw) and Folsom sites.

that had somehow to be dealt with before it spoiled. Mammoth and mastodon jerky may have been staples of Clovis hunting camps in winter. One can imagine a Clovis hunter returning to his tent, proudly announcing the success of his mammoth hunt. Meanwhile, his wife is thinking to herself, "Oh great! Now what do you expect me to do with 500 pounds of meat?"

The Clovis culture was followed by the Folsom culture, identified from the distinctive fluted projectile points first described from a site near Folsom, New Mexico (Fig. 7.5). During the Folsom Period, hunters on the plains of eastern New Mexico began to focus on the hunting of Pleistocene large-horned bison.

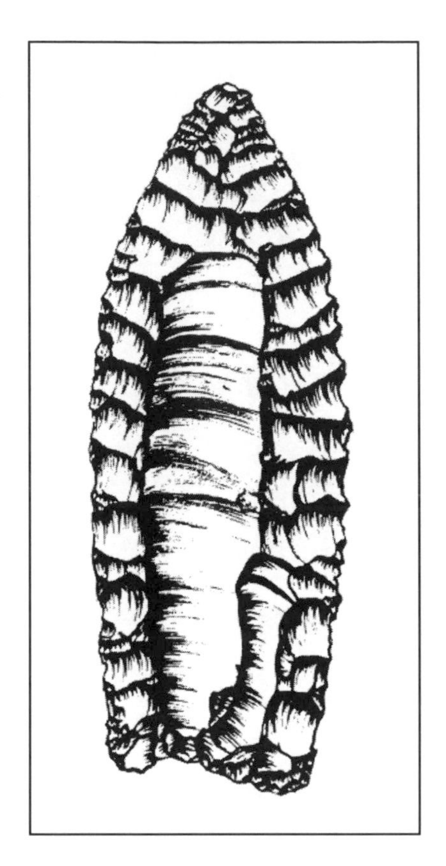

Figure 7.6. Folsom projectile point from the Hanson site, Wyoming. (After Frison, 1991.)

Their projectile points were fluted on both sides; the fluting extended nearly the full length of the point (Fig. 7.6). Based on modern experiments, archaeologists have come to appreciate that this type of projectile point is very difficult to make, requiring careful preparation of a piece of stone, or preform. Because the point is thinned on both sides during its manufacture, breakage is commonplace. Comparisons of the numbers of broken versus intact Folsom points at sites along the Rio Grande Valley showed a failure rate of about 25% in attempts to flute and finish the points. Fluting of the quality seen in Folsom points became a lost art; no younger-age projectile points ever achieved such elegance of form. In fact, Archaic points (from the next younger cultural period) are much more crudely made than either Folsom or Clovis points, in a seeming reversal in technological progress. Other Folsom stone tools included scrapers, bifacial knives, **burins,** and engravers. Tools made from bone included needles, beads, disks, and antler tines from deer and elk, used to flake stone tools.

Along with the bona fide archaeological discoveries in New Mexico, there was one find that has archaeologists scratching their heads in disbelief 60 years later. This was the unearthing of projectile points in Sandia Cave by archaeologist Frank Hibben. Hibben claimed that these points came from an undisturbed layer of sediment, buried under a layer of stalagmite crust and associated with mammoth bones. Based on radiocarbon dating of those bones, "Sandia Man" was thought to have lived in New Mexico more than 20,000 years ago. Several troubling facts have since come to light, however. The layer of sediment that contained the artifacts turned out to be riddled with packrat tunnels. Packrats may have reworked the artifacts into new stratigraphic positions. Uranium-series dates of sediments immediately above the artifact layer revealed that the layer was 300,000 years old. It is difficult to explain how sediments this old were deposited on top of younger artifacts. The mammoth bones and teeth sent to the radiocarbon laboratory in the late 1940s were probably not from Sandia Cave, but from a gravel pit in the Rio Grande Valley that had nothing to do with Sandia Cave. The cave is in the Sandia Mountains, more than 2000 m above the river. So the mammoth remains probably had nothing to do with the human artifacts from the cave. Finally, several archaeologists have examined the projectile points from the Sandia Cave collection and have found evidence that some of them were very recently reworked to match the shape of the others. The Sandia Cave discoveries were big news when they were first published, garnering prominent position in leading scientific journals and archaeological textbooks, but by the 1990s reference to this site has quietly but steadily faded away in the literature. A recent review article by Douglas Preston provides a cautionary tale.

Folsom and Clovis peoples faced large-scale changes in their physical environment, brought about by changing climates and the responses of regional plant and animal communities to those changes. But were these changes at the end of the last glaciation rapid enough to seriously disturb biological communities (including human hunter-gatherers), or were they so gradual that they had little impact on humans? To measure the pace and severity of climate change, it is necessary to look at proxy data, such as insect fossils, that accurately reflect climate and respond rapidly to changing environments. I analyzed late glacial and early Holocene–age insect fossils from a transect of sites along the Rocky Mountain region and south into the Chihuahuan Desert. The insect fossils show a series of rapid, dramatic changes in climate. According to the insect data, as noted in Chapter 6, between 14,000 and 11,000 yr B.P. regional summer temperatures rose about 5°C (9°F). Between 11,000 and 10,000 yr B.P., summer temperatures rose an additional 5°C. The intensity and rapidity of this climatic warming was nothing new in the Pleistocene, however. Evidence from oxygen isotopes and marine microfossils from deep-sea cores indicates that more than a dozen previous glacial-to-interglacial transitions were equally as abrupt and equally intense.

The cumulative effect of such climatic change over a few centuries in a given region may have been one of the primary forces of change in early Paleoindian populations. The intensity of climate change from the late glacial to the early Holocene wrought havoc with biological communities from Montana to Texas, causing regional extirpations of flora and fauna, and possibly extinctions of megafaunal species that were hunted extensively by early Paleoindians. For a given region, the biological "cast of characters" changed quickly, with previously important game species leaving the stage, in some cases never to return. Regional hydrologic changes must have been rather severe in some cases as well, forcing Paleoindian populations to migrate in search of reliable water supplies. This massive hiatus, forcing distributional shifts in species and reshuffling in biological communities, must have been a catalyst for change in Paleoindian ways of life. The insect fossil record—as a sensitive, responsive proxy for climate change—shows how quickly and effectively the "climatic catalyst" was brought to bear in the rapidly changing scene of late glacial to Holocene transition.

How did Paleoindian bands hunt, and how many people were there in a band? These questions are very difficult to answer, based on our very limited knowledge of their culture. Practically the only artifacts preserved from Paleoindian sites in North America are stone tools. We have no direct evidence for how these people made shelters, what clothes they wore, what plants they gathered, and a thousand other details of their daily lives. Some inferences about the numbers of people who hunted together have been drawn from the number of bones of prey animals found together at kill sites. For instance, at a site near Casper, Wyoming, archaeologist George Frison estimated that Paleoindian hunters butchered 42,000 lb (19,000 kg) of usable bison meat. The bison were trapped and killed, apparently by a large band of hunters. Given the amount of meat taken, a large band of people was fed from this kill, not a small, isolated group of only a few families.

Most North American Paleoindian sites have not yet yielded many artifacts other than stone tools. To get a broader view of Paleoindian life-styles, it is necessary to glean information from sites throughout the New World. One such site is Monte Verde in southern South America. This site, because of unusually good preservation of organic remains, offers some fascinating new insights into the way Paleoindians lived in the early postglacial period. Archaeologist Tom Dillehay has excavated this site, situated near the Andes Mountains in Chile. Dillehay found the waterlogged remains of a Paleoindian camp, which had been buried by peat layers and preserved nearly intact. The ancient land surface, preserved under younger layers of sediment, remained essentially undisturbed and showed dramatic evidence of human habitation, including such features as postholes and hearths. The site also contained a wealth of artifacts, which include far more than the typical stone tools. Because of the unusual preservation in a water-

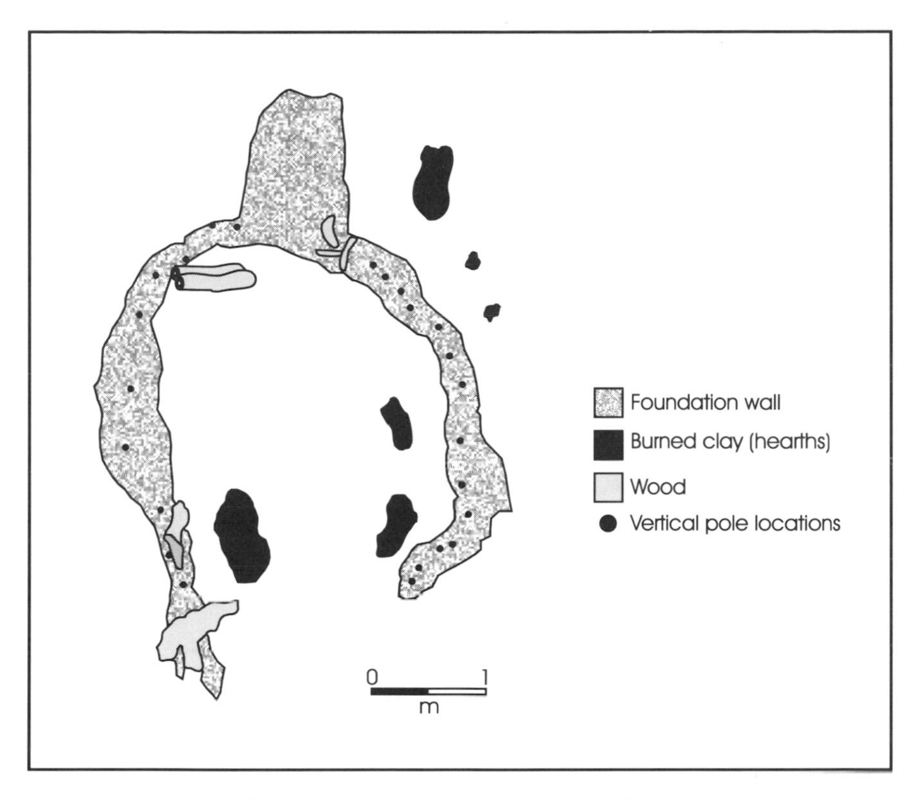

Figure 7.7. Plan view of dwelling foundation at the Monte Verde site, Chile. (After Dillehay, 1991.)

logged environment, the Monte Verde site preserved wooden tools, leather goods, and seeds and other food remains, which have been radiocarbon dated from 13,000 to 12,500 yr B.P. These artifacts predate the earliest reliable Paleoindian sites in North America, including Clovis sites.

The inhabitants of the Paleoindian village at Monte Verde built huts. One hut site was excavated and described in Dillehay's 1986 paper. It had a foundation of compacted gravel, built up outside an earth floor. The walls and ceiling were made of animal hides lashed to wooden poles. Some of the poles and scraps of hide, as well as two hearths and the foundation, were preserved at the site (Fig. 7.7). The excavated hut was part of a group of structures laid out in a rectangular grid of foundational timbers (Fig. 7.8). There were two large hearths outside the dwellings, apparently for community use. This architectural evidence suggests that Monte Verde was a semipermanent village, not just a hastily constructed hunting camp used for only a few days by nomadic hunters.

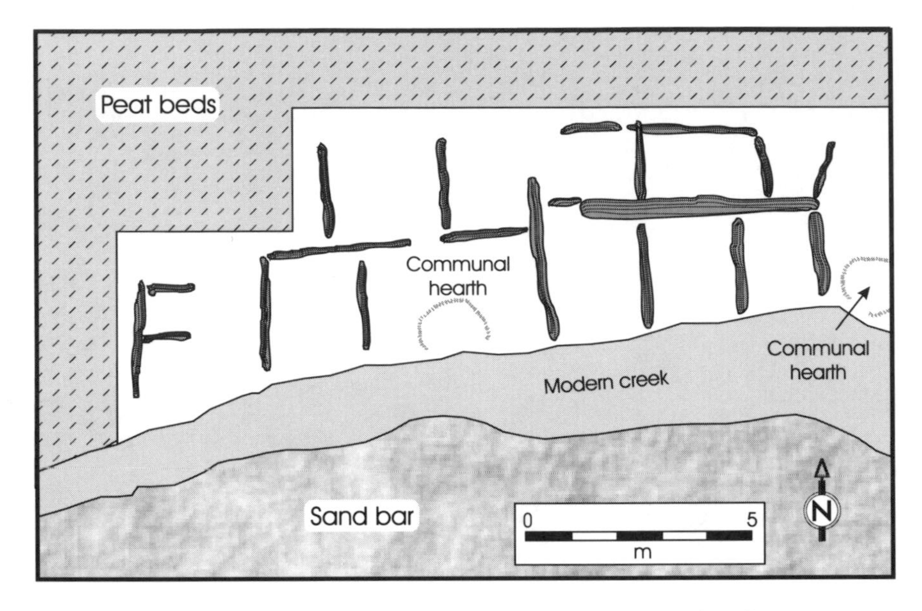

Figure 7.8. Plan view of 12 dwelling foundations along a modern creek at the Monte Verde site. (After Dillehay, 1991.)

Wooden artifacts from the site include spear or dart shafts with stone projectile points mounted on them. The mounting was done by a combination of leather strips and pine pitch (this site provides the first physical evidence of how Paleoindian points were mounted on wooden shafts). Cutting tools were discovered that were made by mounting small, sharp-edged pebbles on wooden handles. Wooden mortars were found, with the remains of ground-up plants and seeds still inside. Several grinding stones were found close by.

Bone and ivory tools were also discovered at Monte Verde. One piece of bone was worked into a shape resembling a projectile point, with a pointed tip and a narrow stem. Mastodon tusk ivory was worked into tools that may have been used as scrapers, based on the patterns of wear on the ends. Bone preservation was excellent at Monte Verde, and the remains of many butchered animals were found, along with mollusk shells. Mollusks were harvested as food, and they may have been an important part of the local diet.

We pick up the trail of the early inhabitants of this continent in a river valley in central Alaska, where archaeologists Chuck Holmes and David Yesner have been excavating the Broken Mammoth site. Here Paleoindians camped, fished, and hunted along the banks of the mighty Tanana River, about 11,500 yr B.P. This site also sheds new light on Paleoindian culture. We have seen that the Monte Verde evidence suggests that Paleoindians were not only big-game hunters. At the Broken

Mammoth site, bones of ducks, geese, and other waterfowl have been found, as well as the bones of large mammals. The site was named "Broken Mammoth" because small tools made of mammoth ivory were discovered there. Salmon bones and scales were also found.

Based on these recent discoveries, and fragmentary evidence pieced together from a suite of sites throughout the New World, many archaeologists have begun to view the Paleoindians as true hunter-gatherers, rather than exclusively as big-game hunters. The relative importance of big and small game in Paleoindian diets is very hard to reconstruct from the archaeological evidence, because it is so dominated by the projectile points used to hunt big game animals. These people simply did not settle down at one site for very long, so they did not leave the sort of refuse piles, human burials, and habitation sites that provide well-rounded evidence of later cultures. In most cases, they left us only a few elements of their stone tool kit and a lot of unanswered questions.

In the Southwest, Clovis hunters hunted a wide variety of prey animals. At Murray Springs, Arizona, and at several sites on the Southern High Plains of Texas, Clovis hunters killed and butchered horses. Smaller animals known to have been taken by Clovis hunters include tapir, peccary, and black bear. Even the mighty short-faced bear was hunted by Clovis people at Lubbock Lake, Texas. Other evidence from Lubbock Lake indicates that the Clovis diet included turtles and turkeys.

Clovis hunters also went after bison in the Southwest. At Murray Springs, Arizona, a bison bone bed lies adjacent to a pile of discarded Clovis projectile points. Paul Martin argues that Clovis hunters wiped out the bison populations of the Southwest. Bison do not live there today, nor did they in late prehistoric times. There is also little or no fossil record of bison bones from Holocene deposits in this region. Lacking competition from other megafaunal mammals in the Holocene, bison flourished on the American Great Plains.

To complete this brief survey of Paleoindian life-styles from throughout the New World, let us turn our attention to a discovery made in Nevada that has shed new light on the technological capabilities of Paleoindians. In 1940, S. M. and Georgia Wheeler, archaeologists working for the State of Nevada, discovered the mummified body of a man in Spirit Cave, a small cave east of the town of Fallon. The mummy was lying on its side, wrapped in a skin robe and sewn into mats woven from tule, a marsh plant. The very dry atmosphere in the cave preserved the mummy, and the mats also helped protect the body. This mummy was initially thought to be at most 2000 years old. It was placed on a shelf at the Nevada State Museum. Recently, researchers at the museum obtained an accelerator mass spectrometer (AMS) radiocarbon date from specimens of hair from the Spirit Cave man, and to their surprise he turned out to be more than 9400 years old. This makes him the oldest known mummy in North America, and one of the oldest in the world. Genetically, the Spirit Cave man appears to be different from modern American Indians. He has a long cranium and a long, narrow face suggesting affinities with Southeast Asian peoples. But perhaps the most remarkable aspect of this discovery are the textiles found associated with the mummy. As I mentioned earlier, almost the only Paleoindian artifacts that have been found in North America are stone tools. Organic remains, such as clothing and other textiles made from plants and animal hides, generally decompose in sediments. But the Spirit Cave mummy and his grave goods were saved from decomposition by the extremely dry air of the cave. The textiles were woven with a method known as diamond plating; they look essentially like modern woven fabrics. As Ronald James, Nevada's state historic preservation officer, remarked, "One bag you could easily accuse them of having bought at a market in Arizona last year." According to an article in the *New York Times,* the woven bags found with the mummy bear a striking resemblance to the open pouchlike African bags that are popular today. The woven textiles found with the Spirit Cave man are easily the oldest known textiles from this part of the world, and the method of weaving was so sophisticated that textile experts are now inferring that Paleoindians must have been weaving fabrics considerably earlier than 9400 yr B.P.

The Archaic period began about 8000 yr B.P., as groups of hunter-gatherer people entered the American Southwest from the west, north, and south. These people foraged for plant foods in addition to hunting both small and large game animals. Their broader food base allowed them to exploit the canyon country of the Colorado Plateau much more successfully than their big-game-hunting predecessors had. In addition to projectile points (Fig. 7.9), they made nets and snares to capture small game, such as rabbits and birds. They also developed stone tools to dig up edible plant roots, cut plant stems, and chop and grind vegetable foods. Their adaptation to life in desert and semidesert regions was quite successful; they spread throughout the Colorado Plateau, and their numbers began to swell. Several hundred archaeological sites that date to this period have been discovered in the American Southwest. Certain similarities in the shape and size of various implements found throughout the region attest to some form of regular trade connections between bands of people.

During the late Archaic Period, plants began to be cultivated for the first time in the American Southwest. At first, people did not depend exclusively on crops for food but continued hunting and gathering. Hunting and gathering had been practiced by all previous cultures in human history, and it was more than just a small step to give up that life-style altogether in favor of farming. Farming is inherently risky, as each year's crop may fail, depending on the vagaries of weather, insect pests, fungal attacks, and other factors beyond human control. Today we

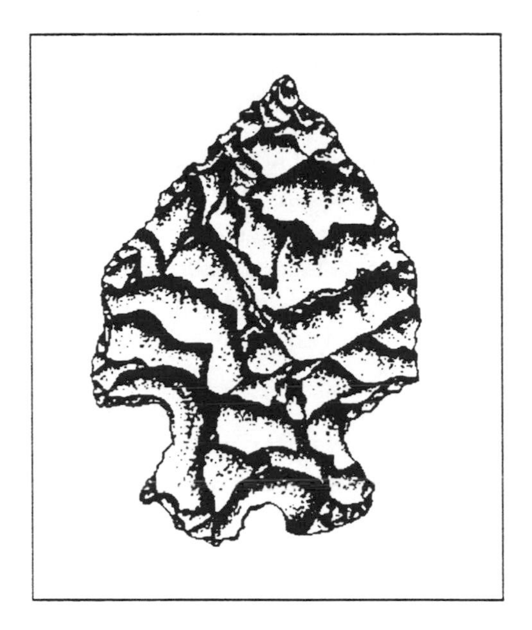

Figure 7.9. Archaic projectile point from the Laddie Creek site, Wyoming. (After Frison, 1991.)

have mechanized farms with chemically sprayed, genetically engineered crops. These modern advantages have taken some, but not all, of the risk out of farming. A single hailstorm can still destroy an entire crop of wheat in Kansas; a summer with too little rain can still shrivel a field of corn in Iowa; a winter with too much rain can still drown a harvest of vegetables in California. Farming has always been a risky business; it was especially risky in its infancy in the American Southwest.

As was the case for the Paleoindian Period, we have pitifully little knowledge of life in the Archaic Period, because hunter-gatherers moved around a good deal, and the remains of their temporary campsites yield little besides stone tools and occasional animal bones. Rockshelters and caves in the Southwest have provided a greater variety of artifacts than other regions, once again because of the better preservation of perishable materials in these dry, protected environments. Cave sites have yielded basketry, sandals made of plant fibers, blankets made of rabbit fur, and twig figurines. Archaeologists have described an Archaic hunting decoy left in a dry cave in Nevada. The decoy was made of reeds, ingeniously woven together into the shape of a duck.

Southwestern Archaic peoples had to cope with environmental changes during the mid-Holocene. Plant macrofossil data developed by Tom Van Devender from southern New Mexico packrat middens indicate that regional climates became more arid after 8000 yr B.P. Julio Betancourt's studies of middens from the Colorado Plateau suggest a shift away from a more abundant moisture pattern there after 6000 yr B.P. In general, desert vegetation expanded throughout the Southwest during the mid-Holocene, at the expense of more mesic woodlands and grasslands. By about 4200 yr B.P., increased precipitation brought higher lake levels to the San Augustin basin of southwestern New Mexico and increased amounts of mesic vegetation to the Chaco Canyon region, as shown in plant remains from a packrat midden sequence from Atl-Atl Cave. In general, the range of climatic variation decreased during late Archaic times, that is, there were fewer large swings in climate. This "settling down" of environmental conditions aided the agricultural revolution that was about to take place. By about 4000 yr B.P., the stage was set for this next step in southwestern cultural evolution—the cultivation of corn.

By about 5000 yr B.P., corn had been domesticated in the Tehuacan Valley of central Mexico (Fig. 7.10). This statement is based on AMS radiocarbon dates from individual corn kernels. Earlier dates (7000 yr B.P.) had been obtained, but they were conventional radiocarbon ages on bulk samples. This is an example of how the AMS method has improved our ability to obtain accurate, reliable ages on small samples.

The development of corn ushered in the single greatest change in North American native life-styles, because people who could cultivate corn year after year in the same place had no need to roam the landscape in search of food. Here, as elsewhere in the world, once people settled down in permanent villages with agricultural fields, the pace of their cultural development increased markedly. By 4500 yr B.P., farmers in central Mexico were growing domesticated varieties of corn, beans, squash, chile peppers, avocado, and amaranth. These food plants were native to the tropical or semitropical regions of Mexico, but more cold-resistant strains of beans, squash, corn, and chiles were developed farther north, and the practice of cultivating these crops eventually spread to the American Southwest.

One of the ways in which the prehistoric development and spread of agricultural products has been traced is through the study of human coprolites (dried feces). Dry caves in central and northern Mexico preserved human coprolites through the last 5000 years of agricultural development, and the coprolites tell the story of diets that shifted from wild, native foods to domesticated animals and plants.

The history of the arrival of agriculture in Anasazi regions is difficult to reconstruct. As far as archaeologists have been able to piece together, it appears that colonists from Mexico brought cultivated crop plants with them into southern Arizona. From there, they either pushed Archaic hunter-gatherers out of the region or greatly influenced their life-style by showing them the advantages of agriculture. Continued introduction of larger and more nourishing strains of corn probably increased its popularity, and the new life-style spread throughout the southwestern region.

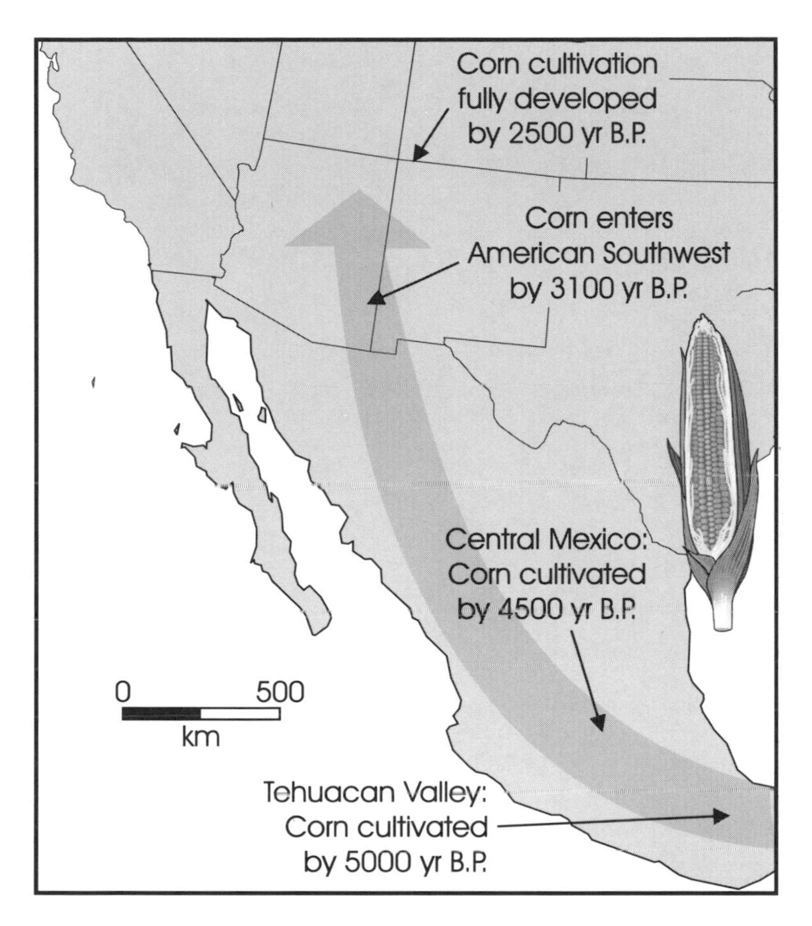

Figure 7.10. Map of southern North America, showing the northward spread of corn cultivation.

The Rise and Fall of the Anasazi

One particular group of prehistoric people in the American Southwest left many ancient monuments, from the fabulous cliff dwellings of Mesa Verde to the impressive ancient cities at Chaco Canyon. They also left us scratching our heads over many questions: Who were these people? What became of them? Why did they abandon their elaborate dwellings? Even the Navajo name we use for them, *Anasazi,* is a bit perplexing. It can mean either "ancient ancestors" or "ancient enemies." The history of the discovery of abandoned Anasazi dwelling places is essentially the history of the archaeology of the American West, since it was the

Figure 7.11. Cliff Palace, Mesa Verde National Park. (Photograph by the author.)

grandeur and mystery of these places that first brought professional archaeologists to this part of the world.

The first Europeans to see Anasazi ruins were Spaniards who traveled through New Mexico and southern Colorado in the sixteenth century. During the 1870s, the U.S. government sent an expedition under Ferdinand Hayden to explore the Four Corners region, recently obtained from the Ute Indian tribe. This expedition included photographer W. H. Jackson and geologist W. H. Holmes. These men documented Anasazi ruins throughout the region (although all of Jackon's photographs were subsequently ruined) and helped popularize southwestern archaeology with the general public in the eastern United States. The group visited several prehistoric pueblos, but it was a more serendipitous discovery by Richard Wetherill and Charles Manson in 1888 that sparked the interest of the entire world.

The Wetherill family owned a ranch near Mancos, Colorado. They were on friendly enough terms with the local Utes to be allowed to graze their cattle in the canyon country flanking Mesa Verde. In the course of their cattle herding, the Wetherills discovered several Anasazi ruins in the region. By far their most exciting discovery came on a cold December day when, in pursuit of lost cattle, Wetherill and Manson stumbled upon an ancient city built in a canyon wall, where rooms were stacked upon rooms in one of the West's first high-rise apartment complexes. There were towers and chambers dug into the ground; the buildings filled the huge cavity underneath the overhanging mesa top. They called it Cliff Palace, and it remains one of the grandest structures ever built by prehistoric peoples in this region (Fig. 7.11). This discovery sparked their interest in the prehistoric peoples of the Four Corners region, and the Wetherills went on to explore the ruins at Chaco Canyon, Canyon de Chelly, and elsewhere. Within a few years, this region had captured the attention of archaeologists from the eastern United States and around the world. In 1906, Mesa Verde became a national park, and the U.S. Congress passed the Antiquities Act, protecting many archaeological artifacts on federal lands from would-be amateur collectors (derisively labeled "pot hunters" by professional archaeologists).

The story of the Anasazi people was not written down, either by the Anasazi themselves or by their contemporaries. However, they left behind a large quantity of physical evidence concerning their way of life, their beliefs, and their connections with other groups in the American Southwest. This unwritten history has slowly been documented through the patient efforts of scientists in a number of fields, largely based on the evidence preserved in the national parks and monuments of this region.

Primitive varieties of corn and squash may have appeared in the American Southwest by late Archaic times, but the first evidence of these plants did not appear there until about 4000 yr B.P. This development marks the transition from Archaic to Basketmaker II culture. It took another 1000 years for the practice of agriculture to become firmly established (Fig. 7.10). Packrat middens from Chaco Canyon and Canyon de Chelly, studied by Julio Betancourt, Tom Van Devender, and Owen Davis, yielded corn pollen in a horizon dated at 3120 yr B.P. and seeds of cultivated squash that dated at 3030 yr B.P.

The agricultural revolution began in earnest about 2500 yr B.P. (500 B.C.), as dated by tree rings from the remains of pit houses and other structures. Agriculture revolutionized life-styles, because for the first time people had the luxury of a stable life in a permanent, settled location. With a reasonable harvest of corn and other crops in the fall, winter no longer had to be a time of living on a knife edge, balanced between starvation and survival.

The Basketmaker III cultural period began about 400 A.D. This cultural horizon is marked in regional archaeological sites by the introduction of indigenous ceramics. The size of Anasazi populations increased during this interval, and people moved to valleys and highlands that were most suitable for farming.

Beans were introduced from Mexico into Anasazi culture during this time. They provided an important source of protein that would otherwise have been lacking in a diet dominated by corn. Together, these two foods provide a well-

Table 7.1. Nutritional Content of Domesticated Crops versus Wild Plants (per 100 g)

	Calories	Protein (g)	Carbohydrate (g)
Crop			
Beans (raw)	340	22.5	61.9
Corn	348	8.9	72.2
Summer squash	19	1.1	4.2
Winter squash	50	1.4	12.4
Wild food			
Seeds			
Amaranth	36	3.5	6.5
Lamb's quarter	43	3.5	6.5
Saguaro cactus	609	16.3	540
Tansy mustard	554	23.4	71.0
Fruits			
Cholla cactus	393	12.2	79.0
Prickly pear cactus	42	0.5	10.9
Nuts			
Black walnuts	606.7	24.3	12.0
Piñon pine	568	11.6	19.4
Other foods			
Cholla stems	—	1.6	—
Mesquite beans	419	14.9	73.0
Purslane	20	1.0	2.6

balanced mix of proteins and carbohydrates. Compared with the wild plants that had been used extensively by Archaic peoples, beans and corn offer substantially greater food value (Table 7.1). Some wild foods, such as walnuts and piñon pine nuts, offer substantial nutritional value, but they are more difficult to harvest in large quantities or are less reliable food sources than crop foods in terms of yearly availability and abundance. Once again, human coprolites provide the information necessary to tell this story.

Parasitologist Karl Reinhard examined coprolites from Mesa Verde and elsewhere. His research shows that Anasazi peoples carried numerous parasites in their bodies, including pinworms, roundworms, threadworms, tapeworms, and possibly hookworms. Today a person would be considered very ill and in need of immediate treatment for such an imposing list of maladies. Yet ancient peoples somehow managed to cope with these parasites on a day-to-day basis. However, it appears that the Anasazi were aware of at least one natural medication to rid themselves of worms. Many Anasazi coprolites contained abundant remains of goosefoot (*Chenopodium*) seeds. This plant is known to have worm-killing properties (it is called an "anthelmintic" in medical terminology). Coprolites with goosefoot remains contained few or no worms. Furthermore, where goosefoot seed was not available, rates of pinworm infestation were much higher. Indeed, the antiworm properties of goosefoot seem to have been widely known in the New World, for an Aztec document (codex) depicts goosefoot as an anthelmintic.

Fossil insects in human coprolites from Mesa Verde show that the Anasazi also had to deal with external parasites. Lice and their eggs have been discovered in the hair of mummified human remains, as well as in hair combs found at Mesa Verde. These mummified corpses were not like Egyptian mummies, which were chemically prepared and wrapped in strips of cloth to ensure their mummification. The Mesa Verde mummies are simply the remains of people who died and were buried in very dry caves. Their soft tissues dried out before decomposition destroyed them.

Interestingly, studies of the coprolites of hunter-gatherer peoples from the Archaic cultural period on the Colorado Plateau turned up no evidence of internal parasites. The parasite problem took hold only when people began congregating in permanent villages and sharing a sedentary life-style. In addition, people walking barefoot over moist soil (as Anasazi farmers certainly did on a regular basis) come in contact with a variety of worms that can gain entrance to the human body by rapidly burrowing through the flesh on the sole of the foot. Forensic studies of human remains from Anasazi sites show that most people died young by our standards (a forty-year-old person would have been considered a senior citizen at Mesa Verde or Chaco Canyon); perhaps constant battles with internal parasites contributed to the brevity of life in prehistoric times.

During Basketmaker III times, permanent dwellings, described by archaeologists as pit houses, were built. The name might suggest a subterranean life-style, but actually the pit house was composed of a floor level excavated just a few centimeters (perhaps 1–2 ft) below the surface, covered by a building rising more than a meter (3 ft) above ground (Fig. 7.12). The aboveground portion of pit houses was made of a framework of poles, plastered over with mud. The remains of several pit houses that date from this period are preserved on top of the plateau at Mesa Verde (see Fig. 7.2). These houses were generally divided into a main room and an antechamber. The antechambers varied in size and shape. A corn grinding area was usually located in one corner of the main room, between the hearth and the antechamber. This area is indicated in many excavated houses by the presence of *metates,* flat grinding stones with a central depression that were used with a *mano,* a cylindrical stone that was scraped over the grain on the metate.

During the Pueblo I period, from A.D. 700 to 900, Anasazi culture became more sophisticated, as expressed in improvements in architecture and ceramics, the development of the ceremonial chamber or *kiva,* and the domestication of cotton.

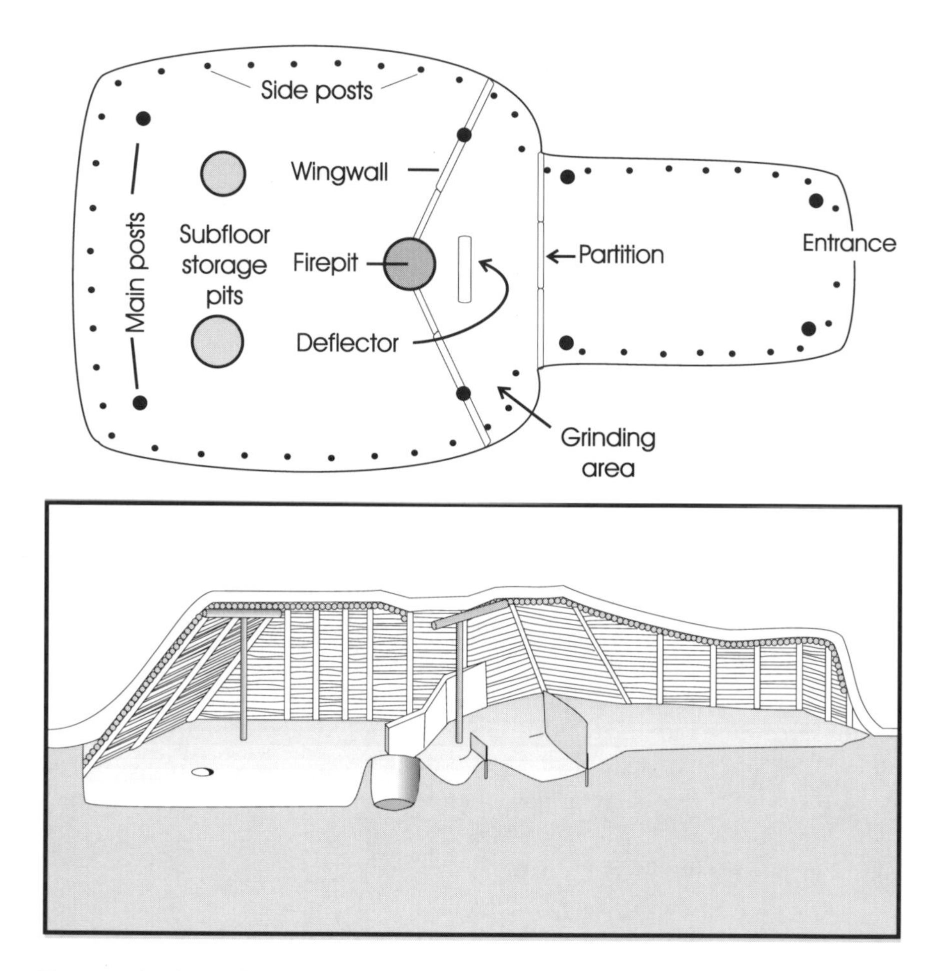

Figure 7.12. Plan and cutaway views of a Basketmaker III pit house. (After Cassells, 1990, used with permission.)

Pit houses were built in rows and groups as communities grew. By the end of this period, pit house exteriors were being finished with masonry rather than mud. This is a logical step in the evolution from the more rudimentary pit houses of Basketmaker III times to the freestanding masonry work used in pueblos and cliff dwellings during the later Pueblo cultural periods.

By Pueblo II times, trade routes between Anasazi settlements had become well established, and a cultural continuity of the Anasazi peoples developed that persisted until the demise of most communities, several hundred years later. It was during this interval that Anasazi numbers grew to their highest levels and the outposts of the Anasazi realm expanded to the edges of the Colorado Plateau and

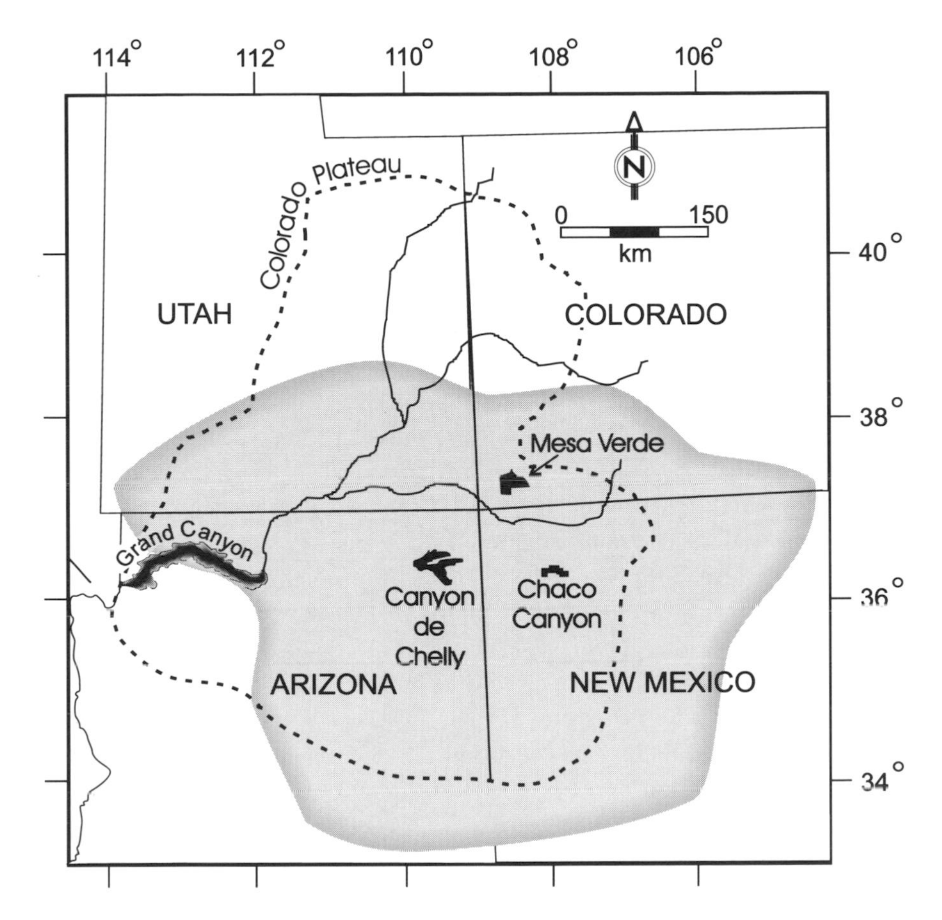

Figure 7.13. Map of the Southwestern United States, showing the extent of Anasazi culture during the Pueblo II stage (shaded region).

beyond. Anasazi settlements ranged west to central Utah and the Virgin River, south to the Mogollon Rim in Arizona, and east to the Rio Grande Valley in New Mexico (Fig. 7.13).

Climatic conditions during Pueblo II times favored agriculture on the Colorado Plateau, and regional farmers began to make better use of the available moisture through the development of field irrigation and water channeling. More productive strains of corn were also developed. The traditional view of the Anasazi at Mesa Verde is that they subsisted almost entirely on a diet of beans, corn, and squash. However, coprolite records that trace the history of human diet through the complete time of occupation at Mesa Verde show that this was not the case. Coprolite evidence indicates that Pueblo II–era diets were dominated by corn, followed in descending order of frequency by prickly pear cactus pads and fruits, squash, meat, pine nuts, and other native plants. The wild food plants that were eaten were species that grew in disturbed habitats, canyon bottoms, and other sites not suitable for agriculture. In fact, squash became less important through time, whereas cotton seed, grass seed, and insects became more important in the diet toward the end of Anasazi occupation.

The importance of corn in the Anasazi diet has recently been documented using two techniques from the paleoecologist's tool kit: palynology of human coprolites and isotope chemistry on bones from Anasazi burials. Palynologist Linda Scott-Cummings studied the pollen content of human coprolites from Mesa Verde and discovered that 95% of the specimens contained corn (*Zea mays*) pollen. However, corn pollen was not the dominant pollen type in the coprolites. Other plants, such as goosefoot, were far more prevalent. Thus, the pollen study left unanswered the question of how much corn the Anasazi actually did consume. The isotope study helped resolve this issue.

The study is based on the chemistry of stable carbon isotopes. Plants use two main chemical pathways in photosynthesis. Most use a process called the C_3 pathway, but others, including corn, utilize the C_4 pathway. The two chemical processes differ in that C_4 plants tend to accumulate CO_2 molecules that contain the ^{13}C isotope of carbon, as opposed to ^{12}C, the most abundant carbon isotope. C_4 plants therefore contain higher amounts of ^{13}C than C_3 plants. As people eat food, they accumulate the carbon (both ^{13}C and ^{12}C) in their tissues, including their bones. Unlike the radiocarbon isotope (^{14}C), these carbon isotopes are very stable and persist in bone tissue long after death. Wild plants in the Mesa Verde region use only the C_3 pathway, so their part of the diet contributes almost no ^{13}C. Because of this characteristic, it is possible to estimate the ratio of C_4 to C_3 plants in the diet of someone who died many centuries ago.

Researchers Kenneth Decker and Larry Tieszen studied rib bones from 35 individuals buried at Mesa Verde, analyzing the ratio of ^{13}C to ^{12}C in the bone samples. The results showed that the amount of corn in the Anasazi diet at Mesa Verde varied widely from one person to another but that its overall importance in the diet remained very constant during the Pueblo period (A.D. 700–1600). The variation in the importance of corn in the diet between different people who lived at the same time indicates that there was a social structure in which some people got more corn and others got less. Nevertheless, this study indicates that corn comprised an average of 70–80% of the Anasazi diet for a 900-year period.

The insect story from Mesa Verde is an intriguing one because it points to changing land use practices. Grasshoppers became an increasingly important element in the Anasazi diet, following the development of agriculture on a large scale. It seems likely that grasshopper populations grew as more soil was cultivated and more crops covered the land. When farmers of any era grow a single species of plant in large quantities, they end up with serious insect pest problems. Pests that are innocuous under wild conditions often increase in number when their habitat becomes a cultivated field, especially when a single type of plant is grown. As any farmer can tell you, grasshoppers love crops. Without any chemical insecticides at their disposal, the Anasazi were probably forced to spend considerable amounts of time physically picking grasshoppers off the leaves of corn, beans, squash, and other cultivated plants. Rather than let this good source of protein go to waste, the Anasazi ate them, and they apparently also sent domesticated turkeys out into their fields to eat insects. This practice yielded double benefits: fewer insects on the crops and fattened turkeys to supplement their meat supply.

Nevertheless, insect pest problems at Mesa Verde may have escalated to the point where croplands on the mesa tops had to be abandoned from time to time. We have no way of knowing if this in fact happened at Mesa Verde, or if it was one of the main causes of the abandonment of the cliff dwellings in the twelfth century, but it is certainly not the least likely scenario to be proposed thus far.

Fossil insect studies carried out by Samuel Graham at Mesa Verde showed that other insect pests presented problems for the Anasazi. Graham studied insect fossils from dried grain stored in pots, from household floors, and from human burials. He found evidence of insect attacks on crops, stored products, and household goods. Evidence from dried corncobs showed that ears of ripening corn had been attacked by a moth species whose caterpillar is called the corn earworm. Crop roots were attacked by a scarab beetle. The adults of this beetle are large, brown chafer beetles with white stripes, and the white grubs (larvae) attack the roots of agricultural crops and other plants. Once the Anasazi harvested their crops, the stored food came under attack from another group of insects: stored-product pests. Some of the same beetle species that are found in grain silos in the United States and Canada today got their start feeding on stored grain in late prehistoric times at such places as Mesa Verde, where relatively large human populations settled down and stored large quantities of grain for the winter. Certain insect pests at Mesa Verde attacked dried corn; others attacked dried beans. Inside Anasazi dwellings, household pests fed on stored food, attacked the dried animal skins and feathers that the Anasazi used for clothing, and lapped up the droppings of turkeys as they wandered through the settlements (it's a disgusting job, but somebody has to do it). The insect scavengers that patrolled the cliff dwellings included larder beetles, darkling beetles, spider beetles, and dung beetles.

One additional line of insect investigation was carried out at Mesa Verde. Graham and his associate Robert Nichols studied the cut timbers that had been used in building construction, looking for traces of bark beetles and other tree-attacking insects. They examined more than 100 timbers of Douglas-fir, lodgepole pine, and piñon pine. All of the timbers revealed evidence of insect attack. Bark beetles burrow under the bark of trees, eating the phloem layer of wood that is in contact with the bark. Based on the pattern of beetle attack, it became clear to the researchers that the Anasazi had cut the trees down in autumn and winter, then left them on the ground until midsummer. By this time, the logs had been thoroughly infested with bark beetles. The bark beetles fed on the logs, loosening the bark to the point that it could easily be stripped off by the Anasazi workers. Thus the Anasazi were letting bark beetles do the otherwise arduous job of stripping the bark off construction timbers. This pattern of behavior became evident because there were no exit holes drilled by the bark beetles in the logs. In late summer or early fall, bark beetles typically cut their way back to the surface of a log, then emerge to find a mate, move to a new tree, and go back inside to lay eggs before winter. But the Anasazi were peeling the loosened bark off the timbers before the bark beetles had emerged, so this removal must have taken place by midsummer (not coincidentally, the most convenient time of the year to build houses, kivas, and other buildings).

Pueblo II times saw the flowering of civilization at Chaco Canyon, described by archaeologists as the "Chaco Phenomenon." I will discuss this aspect of Anasazi culture in the Chaco Canyon section of the chapter. This was a prosperous time for the Anasazi, and they began to move out of pit houses and into small pueblos—compact masonry buildings plastered over with clayey mud (adobe).

The Pueblo III period marked the apex of Anasazi cultural development. The interval from A.D. 1100 to 1300 witnessed the building of the cliff dwellings at Mesa Verde. Overall, fewer settlements were occupied during this interval, as groups of people clustered in a few large communities. This demographic shift anticipated the urbanization of the United States during the twentieth century, but on a much smaller scale. Many of these larger communities, such as that at Mesa Verde, were built on canyon ledges and in shallow caves.

During Pueblo III times, Anasazi diets increased in variety. In addition to corn, squash, and prickly pear cactus pads and fruits, these people also ate amaranth seed, cottonseed, beans, peppergrass and Indian rice grass seeds, and the seeds of several other wild plants. Foraging for edible wild plants undoubtedly increased as cultivated crops failed under drought conditions (or perhaps because of insect infestations) near the end of this period.

By A.D. 1300, Mesa Verde had been abandoned. When the Wetherills and others first explored the ruins, they found many items of everyday use still lying around. The scene gave the appearance that, one day, the leaders of the Anasazi had convinced their people to pack up what they could carry and leave Mesa Verde forever. In fact, the departure was more gradual, but why did the Anasazi abandon a seemingly flourishing community? Several theories have been put forward to explain the decision, most dealing with a change in climate.

Nineteenth-century visitors to Anasazi sites, such as John Wesley Powell, thought that the cliff dwellings had been built as the last retreats of these ancient peoples when their villages on the plains were attacked by Navajos or other raiding bands. Powell and others drew upon the theories of the contemporary Navajos and Utes, who had little direct knowledge of what had happened to the "ancient ones." Today, however, paleoecologists are developing new types of data that help decipher this mystery.

Tree ring data from the Colorado Plateau region, analyzed by Jeffrey Dean from the Laboratory of Tree-Ring Research in Tucson, indicate that droughts occurred throughout this region at A.D. 1150, and again between A.D 1250 and 1450. These periods of diminished precipitation would have led to poorer crop yields that eventually forced the Anasazi to abandon Mesa Verde. The timing of this drought cycle coincides with what climatologists refer to as the Medieval Warm Period or Little Climatic Optimum in Europe. Historical records dating from A.D 900 to 1300. in Europe indicate that this was a time of longer growing seasons (more frost-free days), milder winters, and warmer summers. During this interval, grapes were grown in England (a practice prevented today by cooler climate), and the Norse people founded colonies first in Iceland and then in southern Greenland. The benefits of warmer climate in northern Europe, however, were not experienced around the world. In high-latitude regions, of course, climatic warming benefits human civilization. In arid regions, however, warmer climate, especially if accompanied by drought, spells disaster. It is thought that the same changes in atmospheric circulation that ushered in the Medieval Warm Period in Europe brought drought to the American Southwest.

Prior to the droughts (A.D. 900–1150), advancements in Anasazi culture coincided with the most benign climate of the late Holocene. Water levels in streams were at a late Holocene maximum, precipitation was increasing, and crop yields became more predictable. Climate models indicate that during this period the elevational zone in which upland dry farming (the type of farming carried out by the Anasazi on the mesa tops at Mesa Verde) could be carried out expanded dramatically, giving regional farmers roughly twice the arable land that is available today in the Four Corners region. Unfortunately, this advantageous climate was setting up the Anasazi for the devastating droughts that were to come, because as the population grew so did the people's dependence on good harvests. By about A.D. 1150, large populations were taxing the agricultural capacity of the land. This was also the time when severe droughts began to strike the Colorado Plateau.

Between A.D. 1100 and 1300, the Medieval Warm Period began to break down, to be replaced by the Little Ice Age, a time of intense cold in northern Europe. The onset of the Little Ice Age saw the growth of mountain glaciers in the Rockies, and it brought colder, drier weather to the Colorado Plateau. Tree ring records from

Mesa Verde indicate sharp declines in precipitation from A.D. 1276 to 1299, a twenty-four-year drought. The elevational zone for upland dry farming began to shrink, and it may have disappeared altogether by A.D. 1300. This cold, dry climate persisted in the region until the 1800s, when warmer, wetter conditions finally returned. Before the drought was over, Mesa Verde had been abandoned.

Pollen evidence indicates that after the pueblos were abandoned coniferous trees rebounded on Mesa Verde. It appears likely that this reforestation was made possible because people stopped clearing the land for agriculture and harvesting wood for fuel and construction projects. Some climatic changes may also have contributed to this change in vegetation.

When the cliff dwellings at Mesa Verde were finally abandoned by the Anasazi, they were never again occupied. This fact, combined with the sheltered position of many of the cliff dwellings, contributed greatly to the preservation of the Anasazi buildings at Mesa Verde. When people continue to live in the immediate vicinity of old buildings and monuments they tend to pilfer the building stones and wooden beams for their own use. Thus have ancient edifices been reduced to piles of rubble, including temples in Athens and Rome, medieval castles in Europe, and the buildings of the ancient Aztec capital (now Mexico City). History has been kinder to Mesa Verde. What remains there today provides tantalizing clues to a prehistoric culture that dominated the Colorado Plateau for many centuries, before fading into obscurity. We still do not understand the Anasazi very well, but research tools like the ones used in paleoecology are helping us interpret the unwritten language of their history.

Chaco Canyon: The Flowering of a Civilization

The year was 1849. Lieutenant James Simpson, a surveyor for the U.S. Army, was scouting around northwestern New Mexico when he happened upon a great prehistoric city. He could not believe his eyes, nor could he accept the notion that the ancestors of the Pueblo Indians had built such a magnificent place. Simpson thought that perhaps one of the ancient Mexican civilizations had built a northern outpost; perhaps the Toltecs, or even the Aztecs, had left a monument. In time, other visitors would ascribe Pueblo Bonito (Fig. 7.14), the huge centerpiece of the Chaco Canyon settlements, to the ancient Romans. To be sure, the scale and quality of building, the attention to detail in the masonry, the road system—all these evoke a civilization on a par with that of ancient Rome. Pueblo Bonito still astonishes, still brings many questions to the minds of its visitors. In some ways, the fact that the Anasazi built it is even more remarkable than the supposition that the Romans had built it, for the Anasazi achieved their results without the use of

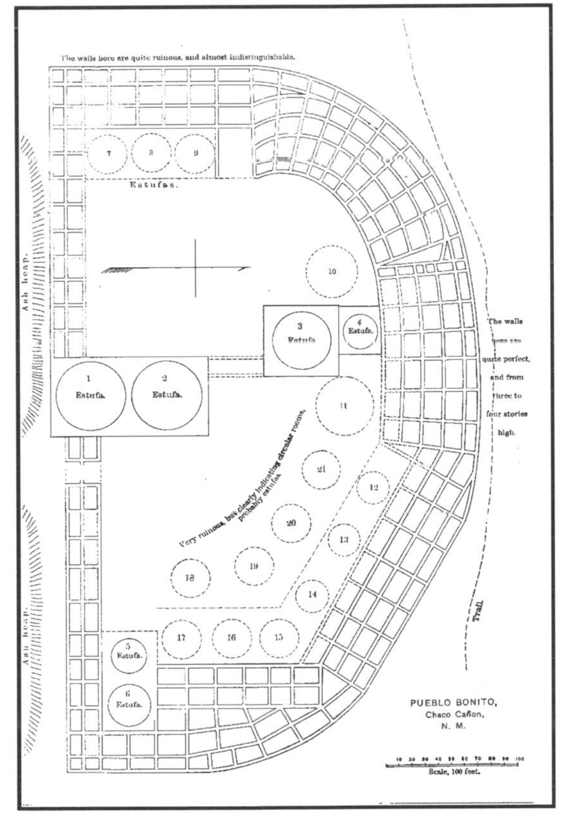

Figure 7.14. Artist's recreation and William H. Jackson's map of Pueblo Bonito, Chaco Canyon, prepared in 1877.

the wagon or other wheeled vehicles, without draft animals to pull the heavy loads, and without a written language with which to pass along engineering and architectural knowledge from one generation to the next.

Other members of that 1849 expedition were scouting around some crevices in a canyon near Chaco when they observed a "dark, pitchy substance agglutinated with the excrement of birds and animals of the rat species." This is the first recorded mention of packrat middens in this region. The soldiers would undoubtedly have been surprised to learn that, more than a century after their expedition, packrat midden evidence would be the key that would solve some of the most important riddles associated with Chaco Canyon.

The grandeur of the Chaco architecture overwhelms the visitor, but it also inspires more questions than it answers. The artifacts and architecture associated with the Chaco pueblos tell us a great deal about the people who fashioned them, but again we are forced to turn elsewhere for answers to some of the most important questions about this place and the ancient people that created it:

1. How did the Anasazi farm here, in such an arid landscape?
2. Where did they get the thousands of large timbers needed to build their pueblos?
3. Why did they abandon Chaco Canyon?
4. What can we learn from the disaster that brought Chaco to an end, and how can we apply that lesson to modern life in the Southwest?

I give here a brief overview of the rise of Anasazi culture at Chaco Canyon, then proceed to delve into paleoecological topics, such as ancient pollen, packrat middens, and tree rings. These subjects may not be as glamorous as the classical archaeology of the canyon, but they provide some of the answers we are seeking.

The rise of Anasazi culture at Chaco Canyon was rapid and dramatic. By A.D. 1100, archaeologists estimate that as many as 5000 people either lived at Chaco or passed through there each year, residing in the many pueblos built along the edges of the canyon (Fig. 7.15). Much of the food necessary to feed these people may have been imported, because the Chaco Canyon region has only enough arable land to feed about 2000 people.

The architectural history of Chaco Canyon follows the same broad outline as that presented previously for Mesa Verde. Basketmaker II villages dating from A.D. 400 to 750 generally consisted of 1–12 pit houses in shapes approximating circular. After 750 A.D., the people of Chaco began building aboveground dwellings. Initially these were made from wooden lattice structures plastered over with adobe. During the Pueblo I period, walls were made of sandstone slabs and adobe, or even crude masonry.

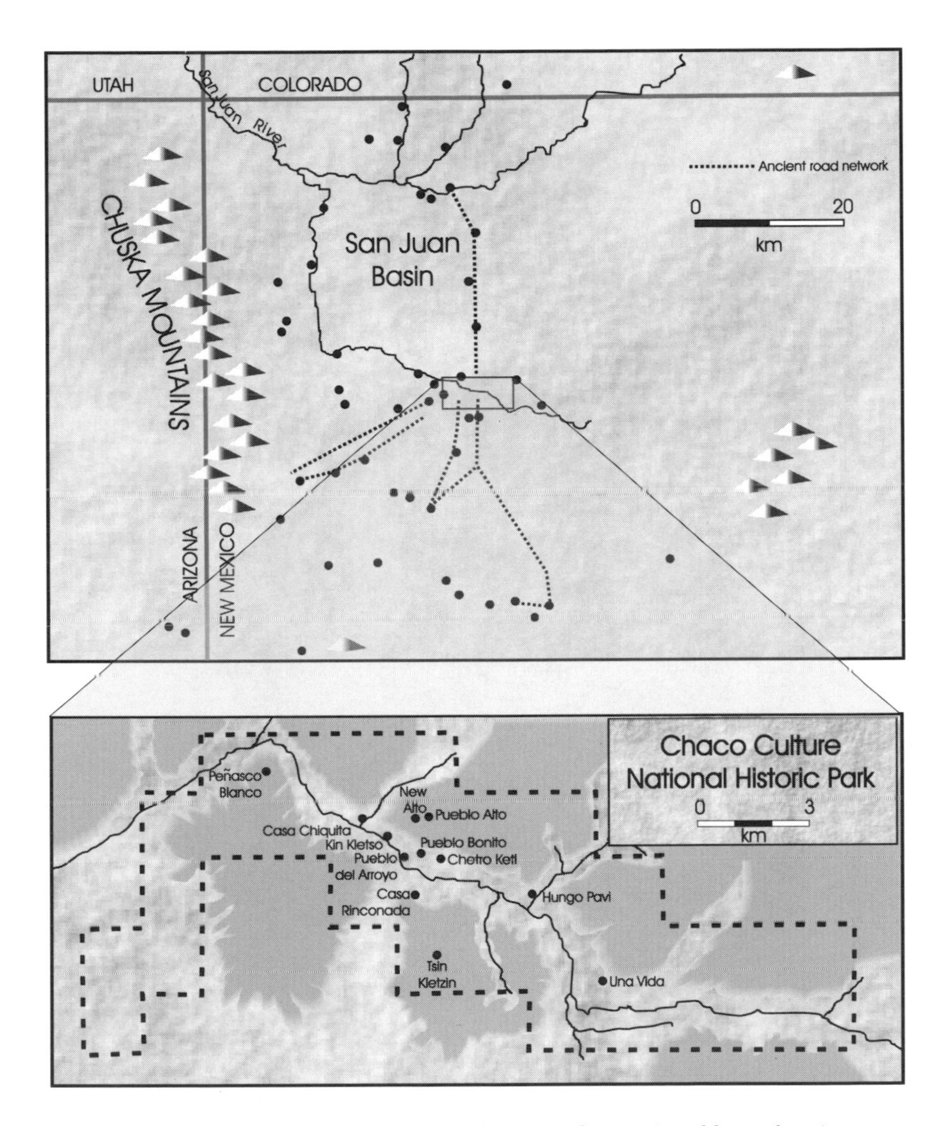

Figure 7.15. Map of Chaco Canyon, showing location of Anasazi pueblos and ancient road network. (After Thomas et al., 1993.)

During Pueblo II times (A.D. 920–1020), the number of rooms per building increased, probably as a result of rising human population levels. Most pueblos continued to be small, built in straight lines with two tiers of living and storage rooms. This style of building was made possible through the use of **coursed masonry** (Fig. 7.16).

The period from 1020 to 1120 is termed the *Classic Bonito* period. During this time, the major town sites were completed, including Peñasco Blanco, Casa Chiquita,

Figure 7.16. Coursed masonry walls, Pueblo Bonito, Chaco Canyon. When occupied, these buildings were plastered over with a layer of adobe. (Photograph by the author.)

Pueblo Alto, Kin Kletso, Pueblo del Arroyo, Pueblo Bonito, Chetro Ketl, Hungo Pavi, Una Vida, and several others (Fig. 7.15). These were magnificent, multistoried structures, made of coursed masonry covered by a layer of adobe. Today the adobe has weathered away, leaving the masonry exposed (Fig. 7.16). The fact that many of the walls remain intact is testimony to the care given to dry-stone masonry construction at Chaco Canyon.

In addition to their elaborate, beautifully made dwellings, the Anasazi at Chaco Canyon also enjoyed an unusually rich material culture. These people had sufficient wealth to afford the best masonry, the best regional pottery, and other expensive goods. One hallmark of wealthy cultures is the quantity of goods thrown away. Poor people cannot afford to waste what little they have, but wealthy people discard much. This was certainly the case at Chaco. At one ancient townsite, Pueblo Alto, archaeologists found 150,000 broken pots. Furthermore, outlying communities such as Canyon de Chelly benefited from technological advancements in such crafts as pottery and masonry (Fig. 7.17).

These people acquired many exotic items through trade with other regions, including shells from California, parrot feathers and copper bells from Mexico, and turquoise from several regions. This trade was facilitated by an extensive system of roads, built by the Anasazi, that connected Chaco Canyon with outlying regions. Satellite imagery has recently revealed the true extent of their road system: it covered over 250,000 km^2 (100,000 square miles), in a series of extremely straight lines radiating out from Chaco.

Chaco outliers (communities closely tied to the Chaco culture) number more than 150 settlements of various sizes. According to Jill Neitzel, an archaeologist at the University of Delaware, these outliers extend well beyond the San Juan basin. It is thought that the people in these outlying communities shared a common ideology and ceremonial life with the Chaco pueblos. The Chaco realm extended northeast to Chimney Rock, Colorado, and southeast to the present town of Guadalupe, New Mexico. Western outliers extended into western Arizona. Chaco outliers did not extend very far east, probably because hunter-gatherers of the Gallina culture kept them out of the Jemez Mountains region. Much of the land to the northeast of Chaco is unsuitable for agriculture, so settlements there would not have fit into the Chaco "system."

A knowledge of the ritual significance of Chaco Canyon to the Anasazi people of a wide region helps us to understand why so many large buildings were built there. Pueblo Bonito is so large and grand not because the local people required such facilities on a day-to-day basis but because it served as a spiritual center for the entire region, where thousands gathered to worship, trade, and celebrate.

At Chaco Canyon, turquoise became more than just an attractive gemstone for use in jewelry. The Anasazi there came to look upon turquoise as something sacred, a bringer of good fortune, a source of spiritual strength. Those who amassed large quantities of the stone were held in great esteem, and the people of Chaco Canyon gained control of the regional turquoise market. The collection and trading of turquoise evolved into a turquoise cult that attracted people from far and wide. If you visit jewelry stores from Santa Fe and Albuquerque to Flagstaff and Phoenix, you will discover that this cult has not yet died out completely. Anasazi families gathered at Chaco Canyon for annual pilgrimage fairs, perhaps to trade for turquoise. This custom brought even greater prosperity to the residents and insulated them somewhat from the ups and downs of annual crop yields; they

Figure 7.17. Glazed ceramic pottery from the Pueblo III period at Canyon de Chelly, Arizona. a–c, Mugs; d–e, jugs; f–i, bowls.

got what they needed, and more, by providing the goods that regional peoples wanted most.

The flourishing of the culture at Chaco Canyon is all the more poignant because it was so terribly brief. Let us now examine the paleoenvironmental record, to track down the circumstances that brought about the demise of this once thriving culture.

The Drought Hypothesis

Tree ring data developed by Jeffrey Dean and others indicate that drought conditions forced the evacuation of Chaco Canyon. Starting in A.D. 1150, a series of droughts ruined crop after crop. These conditions lasted for 50 years. The people of Chaco fought back by building elaborate water-control systems consisting of dams, canals, ditches, reservoirs, terraces, and grid borders around farming plots. The object of these systems was to catch as much of the runoff from the mesa tops as possible, then to channel it to agricultural fields and keep it from running off the fields. Probably more than half of the water used by the Anasazi at Chaco Canyon was runoff from upland streams. Water was diverted to Chaco from increasing distances, but eventually the irrigation system failed, and so then did local agriculture.

Palynologist Steve Hall studied pollen from alluvial sediments that are exposed in arroyos along Chaco Wash and several tributary streams. His research corroborates the hypothesis, based on tree ring data, that during the Basketmaker and Pueblo intervals the climate at Chaco Canyon was drier and warmer than it is today. An interesting footnote to his main paleoenvironmental reconstruction is the decline of piñon pine pollen levels after A.D. 950. Hall speculated that the decline may have been due to harvesting of this species for fuel and construction material during the height of the development of Chaco Canyon (Pueblo III). After Chaco was abandoned, levels of piñon pine pollen began to increase in regional sediments once again. Based on comparisons between the modern pollen fallout in this region and the fossil pollen assemblages, Hall believed that during Basketmaker and Pueblo times the piñon woodland on adjacent Chacra Mesa may have been less than half as extensive as it is today. Hall reasoned that increased erosion of channels used to carry water to the fields may have been the last straw in the collapse of agriculture at Chaco. In a twist on the drought hypothesis, Hall proposed that the cause of the erosion may have been *increased* runoff caused by long-term climate change favoring greater precipitation. A channel, called the Post-Bonito Channel, was eroded down to a level 3.2 m below the Chaco valley floor at A.D. 1100, placing the runoff from upland streams out of the reach of any existing water control system.

The Vanishing Resources Hypothesis

The Chaco residents had another serious problem, one of diminishing resources. As populations grew, the ever-increasing need for wood for fuel and building construction outstripped the capacity of this semidesert region to supply the necessary timber. This is the conclusion of paleobotanists Julio Betancourt and Tom Van Devender, who studied 55 packrat middens in and around Chaco Canyon. At first, they were surprised to find so few remains of piñon pine in Chaco middens. They thought that they might be dealing with samples that were 10,000 years old or older, because that was the last time that piñon was not abundant in this part of New Mexico. Then they started to obtain radiocarbon dates on these samples. It turned out that middens older than A.D. 800 all contained abundant remains of piñon pine. (Packrats are quite fond of juniper and pine boughs, and they bring clippings of these to their nests whenever they are readily available in the immediate vicinity.) It was the midden samples dating from

A.D. 800 and 1500 that were completely lacking in piñon pine macrofossils. This time interval neatly encompasses the period of the greatest Anasazi activity at Chaco Canyon.

It might seem that the droughts documented from A.D. 1150 onward caused the demise of piñon pine at the edge of the woodland, but there were several drought cycles previous to this in the Holocene, and piñon pine remains were found in all of the midden samples that date to these earlier drought intervals. Betancourt and Van Devender drew the conclusion that the people of Chaco Canyon simply used up the available piñon pines, along with junipers, as firewood. The edge of the piñon-juniper woodland had retreated from Chaco Canyon itself to Chacra Mesa, some 20 km to the east.

Michael Samuels and Julio Betancourt developed a computer model of Anasazi fuel consumption. In their model, they assumed that the piñon-juniper woodland around Chaco was relatively sparse, yielding less than 15 cords of wood per hectare (6 cords of wood per acre). The model estimated that, over a 200-year period, the population of Chaco Canyon would have used up all the wood in a 13,000-hectare (32,350-acre) region of piñon-juniper woodland.

Besides using the local trees for fuel, Chaco communities had a large appetite for more exotic types of wood. Wood anatomy studies show that in addition to ponderosa pine, spruce and fir timbers were used in construction. Spruce and fir are high-altitude trees that have not grown anywhere near Chaco Canyon during the Holocene, and the Anasazi had either to travel up to 75 km (45 miles) into regional mountains to get these exotic timbers or to trade for them. An estimated 200,000 trees were used to make the ten major pueblos at Chaco Canyon. The absence of dragging scars on the beams indicates that they were carried to Chaco, not dragged on the ground. Many of these beams were 5 m (16.4 ft) long and weighed upwards of 275 kg (600 lb), so their transport from regional mountain-tops to Chaco Canyon was no small task, especially considering that the Anasazi had no draft animals or wagons. The nearest stands of spruce and fir trees are more than 75 km (46 miles) away, at an elevation of 2450 m (8040 ft).

Although ponderosa pines were more readily available, they were used in such numbers that local populations must have been rapidly depleted. That ponderosa pines were not eliminated altogether is probably due to the fact that the Anasazi carpenters took only trees in a certain size range, leaving other trees behind. This is essentially the same practice modern foresters employ when thinning forests.

Tree ring analyses of felled timbers used in Chaco construction indicate that, unlike the practice at Mesa Verde, tree cutting to meet the needs of Chaco was a year-round occupation, although the majority of trees were felled in spring. The scarcity of stone axes in the ruins at Chaco Canyon lead Betancourt and his associates to an intriguing theory: perhaps there are many dozens of stone axes lying undetected in ancient tree-cutting sites in the surrounding highlands.

The supply of these highland timbers was not easily exhausted; the more pressing problem was the decimation of the local piñon-juniper woodland. By A.D. 1150, the people of Chaco Canyon were out of firewood and very low on water. Their crops were failing. Perhaps the need to import more and more supplies to keep Chaco pueblos going created additional stresses on the people. It was time to move on.

In the end, the Chaco community fled, in search of arable land with more reliable moisture. Some members migrated north to the San Juan River Valley. Others moved southeast, to the Rio Grande Valley. These were the ancestors of the modern Pueblo clans of New Mexico, who were well established in villages along the Rio Grande from Taos to south of Albuquerque by the time the Spanish arrived four centuries later.

But what of the piñon-juniper woodlands that had previously surrounded Chaco Canyon? Betancourt and Van Devender noted that this woodland has not yet recovered from the Anasazi harvesting it suffered 800 years ago. Resource managers and range ecologists might learn a lesson from this story. The southwestern vegetation encountered by Anglos as they entered this region may not have been as pristine as they thought it was. When ranchers cut down large tracts of pinon-juniper woodland to improve forage for cattle, they may be working under the assumption that these woodlands will bounce back quite easily once the cows move on to other lands. The history of the Anasazi at Chaco indicates that this may be a false assumption. In addition, rising public consumption of piñon and juniper wood, especially in regions where the growth of these trees is already marginal, may cause damage that is irreversible on a human time scale.

The lessons of Chaco Canyon and Mesa Verde are clear enough: the landscapes of the Colorado Plateau operate on a very limited water budget. The resources of piñon-juniper woodlands must be used with great care or they will be lost, perhaps for many centuries. When the number of people on the land approaches the ecological limits of sustainability, even a slight change in the physical environment can bring catastrophe. At the end of the twentieth century, our technological advances can forestall some of the dilemmas the Anasazi faced, but for how long?

Suggested Reading

Agenbroad, L. D. 1984. New World mammoth distribution. In Martin, P. S., and Klein, R. G. (eds.), *Quaternary Extinctions: A Prehistoric Revolution.* Tucson: University of Arizona Press, pp. 90–108.

Betancourt, J. L. 1990. Late Quaternary biogeography of the Colorado Plateau. In Betancourt, J. L., Van Devender, T. R., and Martin, P. S. (eds.), *Packrat Middens: The Last 40,000 Years of Biotic Change.* Tucson: University of Arizona Press, pp. 259–293.

Betancourt, J. L., and Davis, O. K. 1984. Packrat middens from Canyon de Chelly, northeastern Arizona: Paleoecological and archaeological implications. *Quaternary Research* 21:56–64.

Betancourt, J., L., Dean, J. S., and Hull, H. M. 1986. Prehistoric long-distance transport of construction beams, Chaco Canyon, New Mexico. *American Antiquity* 51:370–375.

Betancourt, J. L., and Van Devender, T. R. 1981. Holocene vegetation in Chaco Canyon, New Mexico. *Science* 214:656–658.

Cassells, S. 1990. *The Archaeology of Colorado.* Boulder, Colorado: Johnson Books. 325 pp.

Cordell, L. S. 1984. *Prehistory of the Southwest.* New York: Academic Press. 409 pp.

Dean, J. S. 1994. The Medieval Warm Period on the southern Colorado Plateau. *Climatic Change* 26:225–241.

Decker, K. W., and Tieszen, L. L. 1989. Isotopic reconstruction of Mesa Verde diet from Basketmaker III to Pueblo III. *Kiva* 55:33–46.

Dillehay T. D. 1986. The cultural relationships of Monte Verde: A Late Pleistocene settlement site in the sub-antarctic forest of south-central Chile. In A. L. Bryan (ed.), *New Evidence for the Pleistocene Peopling of the Americas.* Orono, Maine: Center for the Study of Early Man, pp. 319–337.

Frison, G. C. 1991. *Prehistoric Hunters of the High Plains,* second edition. New York: Academic Press. 532 pp.

Graham, S. A. 1965. Entomology: An aid in archaeological studies. *American Antiquity Memoirs* 19:167–174.

Hall, S. A. 1977. Late Quaternary sedimentation and paleoecologic history of Chaco Canyon, New Mexico. *Geological Society of America Bulletin* 88:1593–1618.

Lister, R. H., and Lister, F. C. 1983. *Those Who Came Before.* Globe, Arizona: Southwest Parks and Monuments Association. 184 pp.

Petersen, K. L. 1994. Modern and Pleistocene climatic patterns in the west. In Harper, K. T., St. Clair, L. L., Thorne, K. H., and Hess, W. M. (eds.), *Natural History of the Colorado Plateau and Great Basin.* Boulder: University of Colorado Press, pp. 27–54.

Powell, J. W. 1895. *Canyons of the Colorado.* Reprinted in 1961 under the title *The Exploration of the Colorado River and Its Canyons.* New York: Dover Press. 400 pp.

Preston, D. 1995. The mystery of Sandia Cave. *New Yorker,* June 12, 66–83.

Reinhard, K. J. 1988. Cultural ecology of prehistoric parasitism on the Colorado Plateau as evidenced by coprology. *American Journal of Physical Anthropology* 77:355–366.

Samuels, M. L., and Betancourt, J. L. 1982. Modeling the long-term effects of fuelwood harvest on piñon-juniper woodlands. *Environmental Management* 6:505–515.

Thomas, D. H., Miller, J., White, R., Rabokov, P., and Deloria, P. J., 1993. The Native Americans: An Illustrated History. Menomonee Falls, Wisconsin: Turner Publishing. 468 pp.

Yesner, D. R., Holmes, C. E., and Crossen, K. L. 1992. Archeology and paleoecology of the Broken Mammoth Site, central Tanana Valley, interior Alaska, USA. *Current Research in the Pleistocene* 9:53–57.

8

BIG BEND NATIONAL PARK
Life and Times of the Chihuahuan Desert

The Rio Grande begins where small streams catch the runoff from the melting of the abundant snow that falls in the San Juan Mountains of southwestern Colorado. The same mountains that give rise to the Rio Grande are also the source of the San Juan River that heads west when it reaches the southern edge of the San Juan Mountains, running between Chaco Canyon and Mesa Verde to join the Colorado River above the Grand Canyon. So, in a sense, the San Juan Mountains are the knot that holds the various strands of our Southwest story together at the top. These mountains receive a considerable amount of moisture; that water sustains many dry regions to the south and west.

By the time the Rio Grande reaches northern New Mexico, it is a major river, bringing life to the Pueblo villages that dot its banks for the next 300 km (almost 200 miles). The river flows almost straight south through parts of New Mexico then turns to the southeast near El Paso, Texas. It waters large parts of the Chihuahuan Desert from southern New Mexico to southwestern Texas. Where the river cuts canyons through the Sierra del Carmen (Fig. 8.1), a rugged, remote highland region, it changes directions again, turning northeast. This big bend in the river forms the southern boundary of Big Bend National Park, a piece of Chihuahuan Desert and accompanying highlands preserved by the National Park Service since 1944 (Fig. 8.2).

Figure 8.1. The Rio Grande cuts through Boquillas Canyon as it turns northeast to form the Big Bend. (Photograph by the author.)

The park contains a surprising diversity of landscapes and wildlife. Elevations range from 550 m (1800 ft) at the lowest point along the river to 2390 m (7800 ft) at the summit of Emery Peak in the Chisos Mountains. The park represents some of the least disturbed parts of the Chihuahuan Desert. This may seem surprising, given that most of its lowland regions were grazed by cattle as recently as 50 years ago. Sadly, nearly all of the rest of the desert, in northern Mexico, western Texas, and southeastern New Mexico, remains under the hooves of cattle or the farmer's plow. The desert grasslands can support cattle, but only if the size of herds and their grazing range are carefully managed.

The Chisos Mountains are a small volcanic range, the only mountain range contained completely within a national park. They are the southernmost mountains in the United States and are completely surrounded by Chihuahuan Desert lowlands, an island of montane forest in a sea of desert. During the late Pleistocene, woodlands moved down from the mountains onto the plains, giving this region a very different appearance. As in the other southwestern parks discussed in this book, most of our knowledge of ice-age conditions in Big Bend comes from packrat midden records.

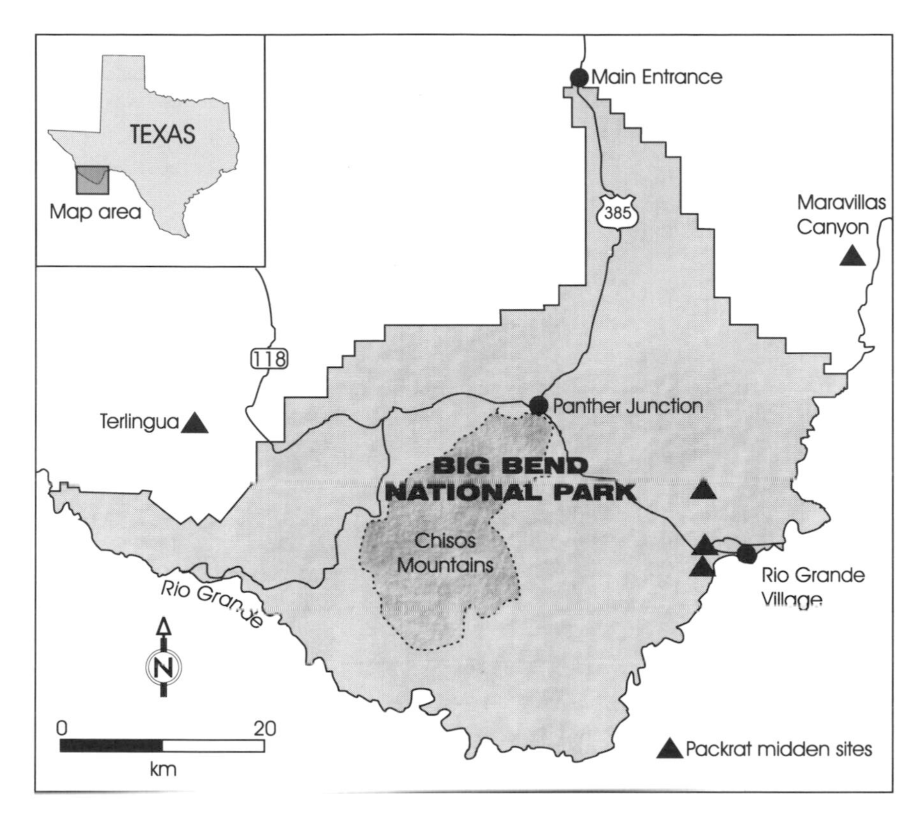

Figure 8.2. Map of Big Bend National Park, showing location of packrat midden sites.

Modern Setting

The Chihuahuan Desert is an interior continental desert, stretching from southeastern Arizona in the northwest to San Luis Potosi, Mexico, in the southeast (Fig. 8.3). The desert lies in the rain shadow of the Sierra Madre Occidental to the west and the Sierra Madre Oriental to the east. The desert comes under the influence of atmospheric circulation from both subtropical regions and regions outside the tropics. In the United States, subfreezing temperatures are not uncommon in winter, and arctic air mass incursions (blue northers) may occasionally bring cold weather to the southernmost parts of the desert. The southern regions support subtropical vegetation, rich in succulent scrub species. These taxa diminish to the north, in favor of more cold-tolerant desert grassland. Creosote bush is an important element in the vegetation in all regions.

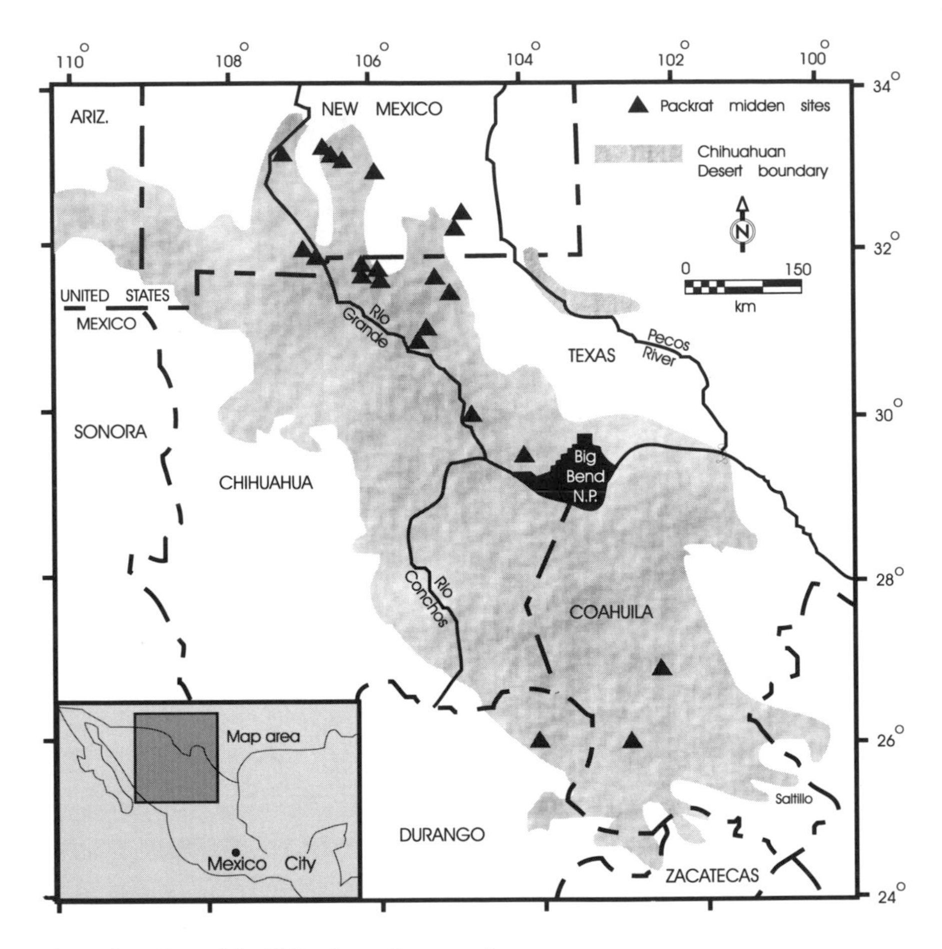

Figure 8.3. Map of the Chihuahuan Desert region.

The vegetation of the park has been divided by ecologist Roland Wauer into the zones shown in Figure 8.4. Beginning at 550 m (1800 ft), the river floodplain vegetation grows in the moist soils. The plants in this riparian zone include fast-growing, broadleaf trees and shrubs, including lanceleaf cottonwood, Texas palo verde, and several species of willow. Giant reeds and common reeds often grow in dense thickets along the river. This well-watered piece of the land forms a very narrow corridor through Chihuahuan desert scrub vegetation (Fig. 8.5). Most of the lowlands in the park are covered in this vegetation type, which can survive on the meager precipitation (less than 250 mm or 10 in. per year) that falls in this region. The desert scrub vegetation zone lies between 550 m (1800 ft) and about 1000 m (3200–3500 ft); this represents roughly half of the park's area. The desert scrub vegetation is rich in species of shrubs, cacti, and other succulent plants, all adapted to soak up as much precipita-

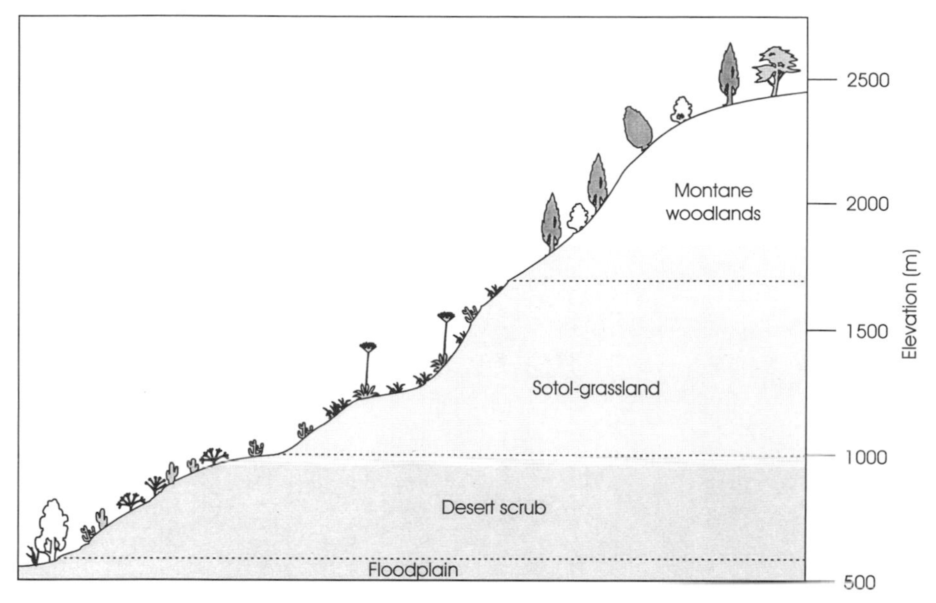

Figure 8.4. Modern vegetation zones of Big Bend National Park.

Figure 8.5. Riparian zone along the Rio Grande at Rio Grande Village. (Photograph by the author.)

Figure 8.6. Lechuguilla in Big Bend National Park. (Photograph by the author.)

tion as possible after the infrequent rains and keep it from evaporating away in the dry desert air. Among the shrubs are several species of legumes that share the characteristic of producing seed pods containing a sugary pulp. These have been important food resources for both prehistoric peoples and wildlife. These plants include the screwbean and honey mesquite. The huisache (tree acacia) and catclaw acacia are other members of the legume family found in Big Bend National Park. Another shrub that yields edible fruit is the guayacan. Its small, heart-shaped pods ripen to expose nutritious seeds. One of the most common plants in the desert scrub vegetation in the park is lechuguilla (Fig. 8.6), a species of the succulent genus *Agave*. Prehistoric peoples of the Chihuahuan Desert made baskets and rope from its fibrous leaves, which also contain cortisone, used then and now as an anti-inflammatory drug.

Candelilla, or wax plant, is a shrub in the widespread desert plant genus *Euphorbia*. As its name suggests, this plant produces a wax. Candelilla has been harvested by the people of the Chihuahuan Desert region for generations. To extract the wax, whole plants are pulled from the ground, then plunged into boiling water with a small quantity of sulfuric acid. The wax comes to the top as a foam that is skimmed off and placed in cooling tanks. The cooled wax is broken into small chunks for transport (traditionally by burro) to towns. It has been refined to make candles, commercial waxes and polishes, phonograph records, and

Figure 8.7. Ocotillo in Big Bend National Park. (Photograph by the author.)

chewing gum. Camps where the wax was extracted abound in the Big Bend region, and a number have been studied as archaeological sites.

Two of the most abundant desert scrub plants in the park are tarbush and creosote bush. Creosote bush grows almost everywhere in the park, in regions below 1200 m (4000 ft). Its pungent odor (especially apparent after a rainstorm) comes from chemicals that are deposited in the leaves as a deterrent to animal browsing. Surprisingly, native peoples have found a use for these seemingly noxious leaves. When boiled to make a weak tea (taken in small quantities at first, then in increasing amounts as a tolerance is acquired), creosote may serve as an anti-inflammatory treatment for such diseases as arthritis.

One of the most fascinating plants of the desert scrub vegetation is ocotillo or coachwhip (Fig. 8.7). During the winter and in times of drought, the tall, spiny stems of this plant look dead. But after a rainstorm, the ocotillo puts out a new crop of small, dark green leaves along its stems. Soon a brilliant cluster of red flowers appears at the top of each stem; the plant is ready to complete its reproductive cycle within a few days of that all-important rain. The long, straight stems of this plant are covered with hard, sharp thorns. When cut and stuck into the ground, the stems take root. The people of the Chihuahuan and Sonoran desert regions have taken advantage of this living barbed wire by planting rows of ocotillo around yards, gardens, and other areas they want to keep protected from livestock (especially goats).

Figure 8.8. Strawberry cactus in Big Bend National Park. (Photograph by the author.)

Big Bend is famous for its many species of cacti. Among these are at least six species of cholla cactus, nine species of prickly pear, strawberry cactus (Fig. 8.8), devil's head, and many others. Cholla cacti have cylindrical, spine-covered stems that may branch into several joints. Prickly pears have flattened joints called pads. Prickly pear cactus pads were an important food to native peoples throughout the arid Southwest. Once the spines are removed (usually singed off over a fire), the pads provide both food value and stored moisture.

The foothill regions above the desert floor in Big Bend comprise almost half of the park area. Sotol-grassland vegetation covers the land at elevations between about 975 m (3200 ft) and 1675 m (5500 ft). As the name implies, this zone is characterized by grasses and by the sotol plant. Sotol is a member of the lily family, as are the several species of yucca that grow here. It has a large cluster of serrated, spiked leaves (hikers traversing the foothills in Big Bend soon learn to wear long pants instead of shorts, no matter how warm the weather) and a large flowering stalk that grows up from the center in the spring. Native peoples relied heavily on sotol for food. The heart, the fleshy base from which the leaves arise (not unlike an artichoke heart), was baked overnight in a mound of dirt, leaves, and rocks that had been heated in a fire. An alcoholic beverage is also made from sotol by boiling the heart and allowing the juices to ferment. Naturalist Roland Wauer describes the taste of this drink as that of a combination of hair oil and gasoline. Botanically the drink has affinities with the Mexican specialty tequila, which is made from the fermented mashed cores of agave plants. Both the agave and sotol store carbohydrates in their fleshy bases.

Other characteristic plants of this vegetation zone include beargrass (not a true grass) and several species of acacia and mimosa, members of the legume family. Notable among these is catclaw acacia, a shrub with curved spines that take hold of passersby and do not easily release their grasp. Attempts to tear yourself away from the curved spines will only bury them more deeply in your clothes or skin.

Elevations above about 1675 m (5500 ft) are clothed in montane woodlands of piñon pine, ponderosa pine, weeping juniper, red-berry juniper, alligator juniper, and Arizona cypress. Weeping juniper, so called because of its drooping foliage, is found nowhere else in the United States. The Chisos Mountains are home to its northernmost population, although the species ranges south through Mexico to Central America. In the Chisos Mountains, Arizona cypress, the tallest of the park's trees, grows only in Boot Canyon and in the canyon above Boot Springs. Ponderosa pine is restricted to just two canyons in the Chisos Mountains. The piñon pine species in the park is Mexican piñon (*Pinus cembroides*).

Broadleaf trees in this zone include aspen and Texas madrone. Aspen grows only on high slopes between the west and north sides of Emory Peak. The bark of the madrone tree peels off in summer, leaving an underbark that turns red when exposed to the air for several weeks. Other broadleaf trees in the Chisos Mountains include mountain maple and nine species of oaks. The oak species are either low-growing trees or shrubs. Most are evergreens.

The wildlife of Big Bend ranges from lizards (Fig. 8.9) and snakes to mountain lion and mule deer. The Sierra del Carmen, on the eastern border of the park, is home to black bear and whitetail deer. Mule deer are generally found at elevations below 1525 m (5000 ft). The park is also home to blacktail jackrabbits (Fig. 8.10), the desert cottontail, ground squirrels, rock squirrels, and two species of kangaroo rat, not to mention packrats (Fig. 8.11). More than a quarter of the park's mammal species are bats, including the cave bat, fringed bat, big free-tailed bat, and sixteen other species.

Canine predators in the park include the coyote, kit fox, and gray fox. The kit fox, a tiny desert fox, lives at elevations above 1050 m (3500 ft). The gray fox lives and roams throughout the park's elevations. Feline species in the park include the bobcat and mountain lion. The rugged mountain country of the Chisos and Sierra del Carmen provides good habitat for the mountain lion, or panther as it is called locally. It gives its name to Panther Junction and a number of other features in the park. A recent wildlife survey counted 21 mountain lions in the park.

No discussion of Big Bend wildlife would be complete without mention of the collared peccary or javelina. This is the only native wild pig in the United States,

Figure 8.9. Long-nosed earless lizard in Big Bend National Park. (Photograph by James C. Halfpenny.)

and it is quite common in the park. These animals are often active at dusk or after dark. If you encounter them on a trail, you are likely to smell them, before you see them, for they give off a strong, musky scent not unlike that of the skunk. They are not likely to see you first, because they have very poor eyesight. However, if you do smell skunk in the park, do not automatically assume the odor comes from a javelina: four species of skunks live in the park as well. While mammalogist James Halfpenny and I were live-trapping for packrats a few miles outside the park, we managed to catch a spotted skunk. The frightened, angry animal just about filled the cage we had caught it in, and we took considerable time and trouble to release this unwanted captive from its confinement!

Among the many species of reptiles found in the park are six species of poisonous snakes. These include the prairie rattlesnake (possibly wiped out by now), the black-tailed rattlesnake, the western diamondback rattlesnake, the Mojave rattlesnake, the mottled rock rattlesnake, and the Trans-Pecos copperhead. The rattlesnakes frequent a wide variety of habitats from the banks of the Rio Grande to the Chisos Mountains. The copperhead is most common in patches of cane and other sheltered locations along the river. Some visitors are spooked by the very thought of these snakes. Actually, if you hike around the park, you are more likely to tangle with a catclaw acacia than with one of these snakes. But if you do see one, back off.

The snakes are most active in the heat of the summer. At that time of the year it is far more pleasant to hike in the cool of the early morning, and this is also the time of day when the snakes are least likely to cause you any trouble, since they must warm themselves before they become active.

Shifting Life Zones in the Late Pleistocene and Holocene

To understand the history of Big Bend National Park, it is necessary to fit it into the context of events throughout the larger Chihuahuan Desert region. Dozens of packrat midden sampling sites have been studied throughout this desert, from southern New Mexico to the Mexican states of Chihuahua, Coahuila, Durango, and Zacatecas. The midden samples studied from this desert range in age from >43,300 yr B.P. to recent.

The paleoenvironmental record developed from packrat middens shows that this desert experienced a series of environmental changes through the late Pleistocene and the Holocene. These modifications brought about large changes in the distributions of Chihuahuan plants and animals. Some of these species movements were from one part of the desert to another; other movements were outside

Figure 8.10. Blacktail jackrabbit in Big Bend National Park. (Photograph by the James C. Halfpenny.)

Figure 8.11. Packrat in Big Bend National Park. (Photograph by James C. Halfpenny.)

its boundaries (i.e., regional extirpations and establishment of species new to the region). Curiously, even though vegetation and insect communities shifted about a good deal at the end of the Pleistocene, the small vertebrate fauna of the Pleistocene (e.g., rodents, rabbits, lizards, and snakes) was essentially the same as the modern fauna. In other words, the small animals experienced few changes in their distribution through this series of environmental changes.

In the northern part of the desert (southern New Mexico and northwest Texas), piñon-juniper-oak woodland dominated limestone slopes during the middle to late glacial interval (42,000–10,800 yr B.P.). During the early Holocene, an oak-juniper woodland community developed. This gave way to desert grassland after about 8250 yr B.P. This desert grassland probably had the general appearance of the modern desert scrub vegetation, but the mixture of species was somewhat different than that found today.

By about 4200 yr B.P., the common elements of the Chihuahuan Desert flora had migrated into the northern Chihuahuan Desert region, forming a relatively modern desert scrub vegetation on rocky slopes. This vegetation remains the dominant one in the region today, and during the nineteenth and twentieth centuries plants such as honey mesquite, prickly pear cacti, creosote bush, ocotillo, and other shrubs have invaded much of the grassland regions. This invasion was most likely brought on by large-scale disturbances of the landscape, including cattle ranching and the prevention of range fires. Cattle tend to disperse mesquite by eating the sweet pods and then voiding the seeds; they avoid noxious plants such as creosote bush. Thus by one means or another these species are favored on the landscape, especially when grasses are grazed to the extent that they can no longer effectively compete with shrubby species. Fire prevention also favors the shrubs. Natural fires burn these shrubs and cause many to die, whereas grasses come back even more successfully after range fires.

In the central part of the Chihuahuan Desert, in and around Big Bend National Park, the vegetation during the middle of the last glaciation (40,000–22,000 yr B.P.) was a piñon-juniper-oak woodland on the uplands and juniper **parkland** on the lowlands. The vegetation under and between the trees was dominated by grasses and other herbs.

By late glacial times (after 21,000 yr B.P.), papershell piñon became important in the upland flora of Big Bend, and piñon-juniper-oak woodland spread into lowland regions, indicating relatively mild winters, cool summers, and more precipitation than today. The boundary between late Pleistocene and early Holocene environments (between 11,200 and 10,300 yr B.P.) is well marked in the vegetation, with the demise of papershell piñon and the increase of scrub oak before 8200 yr B.P.

Succulent desert scrub vegetation developed in Big Bend during the early Holocene. This vegetation included some of the familiar plants from the modern landscape, such as lechuguilla, catclaw acacia, honey mesquite, and prickly pear cactus. Vegetation diversity increased in the mid-Holocene, indicating a switch to subtropical climatic conditions. Middle Holocene vegetation in the lowlands of Big Bend was essentially modern Chihuahuan desert scrub. Lowland midden samples from 1700–1600 yr B.P. contained a very dry-adapted (**xeric**) flora, and the modern vegetation is as xeric as any in the last 10,000 years.

In the southern Chihuahuan Desert of Mexico, the late glacial vegetation on limestone slopes (where the packrat middens are found) was a woodland dominated by juniper and papershell piñon in association with succulents such as sotol and beaked yucca. Piñon pine abundance declined after 13,000 yr B.P., and the number of succulents increased. At some time between 11,730 and 9360 yr B.P. the regional vegetation shifted to a Chihuahuan desert scrub. Apparently, no transitional woodland formed during the shift from the piñon-juniper community to desert scrub. The fossil plant record from this region shows unique combinations of species from various modern regions and habitats. Late glacial plant assemblages included both conifers and desert succulents, but temperate plants dropped out of regional records during the middle and late Holocene. A mixture of temperate and dry habitats has existed in this region, even during the last 1000 years, when other regions of the Chihuahuan Desert have experienced extremes of aridity.

Figure 8.12. Maravillas Canyon Cave, Black Gap Wildlife Management Area. (Photograph by the author.)

The reconstruction of Pleistocene vegetation in and around Big Bend National Park began with the work of Phillip Wells in the 1960s. Wells pioneered the use of packrat middens in southwestern paleoenvironmental studies, and Big Bend was one of the first regions to be studied. In the 1970s and 1980s, paleobotanist Tom Van Devender of the Arizona-Sonora Desert Museum worked on packrat middens from sites throughout the Chihuahuan Desert. Tom's work has focused on packrat middens at several sites in the Big Bend region (Fig. 8.2). One of the most important sets of fossil records comes from Maravillas Canyon Cave. This cave (Figs. 8.12 and 8.13), situated along a canyon in the Black Gap Wildlife Management Area, produced 14 midden samples, recording the vegetation history from 27,800 to 1200 yr B.P. The canyon is formed by a major tributary of the Rio Grande just east of Big Bend National Park. The vegetation surrounding this low-elevation cave is not as dry adapted as the lowlands in the park, owing to its position near the northeastern edge of the Chihuahuan Desert and its north-facing aspect.

The paleobotanical record from Maravillas Canyon shows mid-Wisconsin glacial vegetation of piñon-juniper-oak woodland. A woodland with more papershell piñon replaced this vegetation prior to 21,000 yr B.P.

The boundary between mid-Holocene and late Holocene vegetation fell after 4700 yr B.P. This transition is marked by an increase in mesic vegetation communities, culminating in the last 1000 years. Unfortunately, there are too few species preserved in the Holocene insect fossil record from Maravillas Canyon to allow many direct comparisons in this interval.

The second major study area for the park region includes packrat middens from several sites in the Rio Grande Village area (Fig. 8.2). These sites document the vegetation history of the last 36,000 years, based on 35 packrat middens from caves and rockshelters (Fig. 8.14). This is the lowest, hottest, and driest region in the modern Chihuahuan Desert. Falling in the rain shadow of the Sierra del Carmen, the vegetation of the Rio Grande Village area has been drier than that of Maravillas Canyon for at least 28,000 years, but changes in precipitation were reversed during the Holocene, causing Rio Grande Village to be wetter when Maravillas Canyon was drier. An additional site near the town of Terlingua lies just west of the park boundary. One sample, dated at 15,000 yr B.P., provides a late glacial macrofossil record of papershell piñon, juniper, and shrub oak. According to Van Devender, this region supported the most xeric woodland community recorded for the late Wisconsin from the Chihuahuan Desert.

The history of creosote bush in the Chihuahuan Desert provides a good example of how the vegetation of this region is constantly shifting in response to environmental change. The oldest records of creosote bush are somewhat tenta-

Figure 8.13. Close-up of the inside of Maravillas Canyon Cave, showing packrat middens (dark deposits at the back of the cave). (Photograph by the author.)

Figure 8.14. The Tunnel View packrat midden site, near Rio Grande Village, Big Bend National Park. (Photograph by the author.)

tive, isolated remains in middens from the Rio Grande Village region, dated at 26,430 and 20,450 yr B.P. Some biogeographers have postulated that creosote bush spent the Pleistocene in localities south of the Chihuahuan Desert in Mexico, but Van Devender believes that this plant may have survived the late Pleistocene in the Big Bend region, in small populations growing in particularly xeric habitats.

The postglacial history of creosote bush in the Chihuahuan Desert tracks the spread of arid climates from south to north through the Holocene. Creosote bush was found in Rio Grande Village middens dated at 10,000 yr B.P. This date is an accelerator mass spectrometer date on the creosote bush remains themselves, so there is no question of contamination from younger specimens. Rio Grande Village is one of the lowest, hottest places in the Chihuahuan Desert today, and it probably has been that way throughout the late Quaternary. By 8200 yr B.P., creosote bush had become established at Maravillas Canyon, but it did not reach the northern parts of the Chihuahuan Desert region until about 3500 yr B.P. The northernmost populations of creosote bush have only become established within the last few centuries. It now grows as far north as Isleta Pueblo, New Mexico. Cattle grazing appears to favor cresosote bush at the expense of grasses and herbs, so human influence has helped the spread of creosote bush in regions where it otherwise might not flourish. Cattle will not eat this plant, as the leaves contain some noxious compounds that deter herbivores. In fact, packrats are among the few mammals that do not seem to be bothered by the chemicals in creosote bush leaves. Given the contribution that packrats have made to our understanding of Chihuahuan Desert paleobotany, it seems only fitting that they have the last word on this subject.

Fossil Insect Records

In the last 10 years, insect fossils from packrat middens have received intensive study. Although fossil insect research is now underway at sites in the Sonoran Desert and the Great Basin, the Chihuahuan Desert has been the most intensively studied region. I have studied more than 200 insect fossil samples from middens at 35 sites, ranging from the southern part of the desert region in Mexico to the northernmost part of the desert in central New Mexico. The fossil insect data are rich, with more than 200 identified species of beetles, ants, and bugs, as well as other arthropod groups such as spiders, scorpions, and millipedes. These fossils provide many new insights into the timing and intensity of environmental changes in this region.

Late Pleistocene insect faunas were mixtures of temperate and desert species, combinations of species not seen in any one region today. Since the end of the Pleistocene, some of these species have become established in different regions of the Chihuahuan Desert. Others now live outside this desert. The fossil insect record indicates that even sedentary, flightless beetles (such as heavy-bodied weevils) have undergone marked distributional shifts in the Southwest. Whereas shifts in plant distributions have taken up to a thousand years or more to be accomplished, the changes in insect distribution took place within the space of a few decades to centuries. Moreover, even highly specialized cave dwellers have somehow managed to move from one cave system to another in response to changes in late glacial and Holocene environments.

The regional reconstructions are as follows. The fossil insect assemblages from the southern Chihuahuan Desert contain mixtures of desert-dwelling and temperate-zone species in almost every sample from late glacial times through the late Holocene. Midden assemblages from locations farther north in the Chihuahuan Desert are generally separated into glacial-age faunas dominated by temperate-zone species and Holocene faunas dominated by desert-dwelling species. The "no-modern-analogue" faunal assemblages (mixtures of species that do not occur together in any one place today) indicate that the late Quaternary environments in this part of the desert were unlike any that exist today. This is the same conclusion that has been drawn from the plant macrofossil record.

The insect faunas from the Big Bend region suggest greater effective moisture from 30,000 to 12,000 yr B.P. Many temperate grassland species lived in the Big Bend region during late glacial times. After 12,000 yr B.P., most of these species were replaced by either desert species or more wide-ranging taxa. This change in insect inhabitants suggests that the climate shifted from the cool, moist conditions of the late glacial to the hotter, drier conditions of postglacial times.

During the height of the last glaciation (22,000–18,000 yr B.P.), the northern Chihuahuan Desert arthropod records suggest widespread coniferous woodland at elevations as low as 1200–1400 m (3950–4600 ft). These woodland environments lasted until about 11,000 yr B.P., but the insect data suggest that considerable open ground existed, and grasses were well developed at the midden sites. The grassland signature of the arthropod fauna was also suggested in the regional vertebrate record. The transition from the temperate glacial fauna to the more xeric Holocene fauna started by 12,500 yr B.P. The timing of this faunal change was essentially the same throughout the northern Chihuahuan Desert.

A major difference between the Big Bend and northern Chihuahuan Desert scenarios is the nature of this insect faunal change at the end of the last glaciation. In the Big Bend region, the transition was characterized by the disappearance of all but one of the temperate insect species at about 12,000 yr B.P. However, the desert-dwelling insect fauna did not appear in the Big Bend records until about 7500 yr B.P. In between these times, a more cosmopolitan fauna existed there. Unfortunately, this fauna provides little paleoenvironmental data. In the northern Chihuahuan Desert insect assemblages, the desert-dwelling species first appeared at 12,500 yr B.P., and several of the temperate grassland species from the Wisconsin interval persisted well into the Holocene. This mixture of desert and grassland elements makes sense from an ecological standpoint, because these northern faunas were living close to the edge of the Chihuahuan Desert. During the Holocene, the gradual shifting of the northern desert boundary probably created many marginal habitats for temperate species in ecotones between grassland and desert scrub communities.

By about 7500 yr B.P., the appearance of more xeric species indicates the establishment of desert environments, including desert grasslands, throughout the Chihuahuan Desert region. After 2500 yr B.P., the last of the temperate species was replaced by species associated with desert scrub communities.

Tom Van Devender and Eugene Hall have recently begun to study insect assemblages from packrat middens in the Sonoran Desert. The comparison between these faunas and the Chihuahuan Desert faunas is instructive, because it shows that the two regions had quite different environments during the last glaciation. Unlike the Chihuahuan Desert insect fauna, the composition of Sonoran Desert insect communities underwent little change between late glacial and modern times. All the species found in the Sonoran midden fossil assemblages live within a short distance of the midden sites today. The Sonoran fauna showed a marked increase in diversity during the late Holocene, as a more subtropical climate became established about 4000 yr B.P. Many warm-adapted insects probably migrated into southern Arizona from northwestern Mexico during the last 4000 years. This late Holocene peak in species richness contrasts sharply with the Chihuahuan Desert insect record, in which the least number of species appear in late Holocene samples. If the Sonoran Desert insect fauna truly was stable through the late Quaternary, then this represents a significant difference between the Sonoran and Chihuahuan Desert insect faunal histories.

Probably the most interesting questions concerning the fossil insect species from the Chihuahuan Desert assemblages are the following: Where are they now? What has become of the late Pleistocene insect fauna of this region? There is no single answer. Many species are still in the Chihuahuan Desert, though they may have shifted their distributions to different locations. Others have departed for other regions. In fact, the various species that have departed have gone in every possible direction. Several species have shifted westward, into the Sonoran, Mojave, and Great Basin deserts. Some species have shifted eastward onto the plains of central Texas. Others have moved to the northeast, and others have shifted to the north. Finally, a few species have shifted southward in Mexico. Furthermore, the timing of distributional changes within the Chihuahuan Desert and emigrations out of that desert follows no single pattern. Some species began their departure from the Chihuahuan region as early as 15,000 yr B.P. Some temperate-zone species began shifting north from the southern part of the Chihuahuan Desert by 12,500 yr B.P. but remained in the northern desert region until the early Holocene; at least one temperate species survived in the hills of south-central New Mexico until 1500 yr B.P. Today this beetle is found only in the grasslands of the Great Plains region. Other species began their departure from the Chihuahuan Desert only in the Holocene.

Some of the most puzzling distributional changes of Pleistocene Chihuahuan Desert beetle species are those of a group of cave-dwelling ground beetles. These beetles are flattened top to bottom, enabling them to maneuver through narrow cracks and crevices in caves. They are flightless and pale white, having no pigmentation; most have reduced eyes or are blind (eyes are useless in the perpetual darkness of deep caves). The mystery of the changes in distribution lies in their ability to move from caves in the Chihuahuan Desert to caves in central Texas and Oklahoma. There are no known subterranean connections between these far-distant cave systems, so how did they travel from one cave system to another? One such species is known today only from Carlsbad Caverns, New Mexico, and nearby caves within Carlsbad Caverns National Park. Based on the modern evidence alone, one might suppose that this beetle has never strayed from the caves of the Carlsbad region, but I found fossil specimens from as far away as Big Bend National Park. Some modern biologists have suggested that the ancestral stocks of these cave beetles have been isolated in their current cave systems for several million years, but the fossil evidence shows that these species are able to migrate substantial distances within at most a few thousand years. This is a good example of the folly of trying to figure out the history of a species, armed only with modern biological data. The fossil record is the only reliable source for such information. It cannot be anticipated or mediated. It must be unearthed!

Early Peoples of the Big Bend Region

The Chihuahuan Desert has been a marginal region for human populations since the earliest Paleoindians arrived in the Southwest just after 12,000 yr B.P. The desert probably never supported large herds of prey animals, even in late glacial times. However, signs of human occupation throughout the Holocene are fairly common in and around Big Bend National Park. Paleoindians of the Clovis and Folsom cultures were probably casual visitors to this region. They enjoyed considerable success in hunting big game animals a few hundred kilometers to the north, on the Southern High Plains of western Texas and eastern New Mexico. Archaic Period peoples used the park region extensively. These hunter-gatherers trapped small game and foraged for edible plants. It is likely that they used the Big Bend region on a seasonal basis. They probably made this region part of their annual circuit, exploiting the surges in wildlife and food plant productivity that are tied to the annual precipitation cycle.

About 1000 yr B.P., three new technologies were introduced into the Big Bend region: ceramics, agriculture, and the bow and arrow. Pottery production and farming took place only along the Rio Grande valley, where access to a reliable water source was not a problem. The residents of the desert regions were still hunter-gatherers. These were the people encountered by the first Spaniards who entered the Big Bend region, soon after Hernando Cortéz and his men overthrew the Aztec Empire in Tenochtitlán (modern-day Mexico City) in 1521. At that time, the desert Indian tribes of northern Mexico and southwestern Texas included the Tobosas, Saliñeros, Tepehuanes, and Chisos. The farming tribes along the Rio Grande were the Jumanos and Patarabueyes. These groups were influenced by the Pueblo farmers to the north, who had villages as far south as El Paso. The Spanish policy toward the natives of this northern frontier region was to subdue the people by military conquest, then impose their civilization upon them. They also pitted one group against another to weaken both. For instance, the Spanish employed the

Saliñeros to fight their old enemies, the Tobosas and Chisos. Sadly, this subjugation and exploitation led to the rapid decline of native populations.

By 1690, new tribes, including the Apaches and Janos, had come down from the north to raid Spanish settlements. Desert military outposts were set up throughout northern Mexico and southwestern Texas to defend against these tribes. The influence and regional control of various tribes waxed and waned for the next few decades, but by the mid-1700s nearly all of the original inhabitants of this region had been exterminated. By the 1740s, Apaches had taken over the entire region, expelling the Spaniards. The last sixteen members of the Chisos tribe were imprisoned in Durango, Mexico, in 1748. In 1744, the Comanche tribe began raiding in the Big Bend region. For more than a century, the Spanish, Apaches, and Comanches fought for control of the northern Chihuahuan Desert region. Indian raids continued well into the ranching era in the 1890s.

It seems strange that so many groups of people should be willing to fight for such a land as this, with its harsh environment and limited resources. But fight they did. When the Indians stopped raiding, Mexican bandits took up the slack, rustling cattle, robbing ranches and towns, and escaping back across the Rio Grande. One of the most infamous of these was Francisco Villa, better known as Pancho Villa. The bullets and arrows only stopped flying after the Texas Rangers and U.S. Cavalry took charge of the region, establishing law and order by 1919.

Big Bend is an exhilarating landscape because of its great diversity. The terrain varies from the floodplains and canyons carved by the river to desert flats and foothills to mountain crags. These landscapes support a wide variety of plant and animal communities, a microcosm of a much broader transect of the continent encapsulated into one highly variable region. The severity of the summer heat tests every species; each has its own way of coping with the desert conditions. All the spines and barbs on the plants are not there just for protection. They are also an adaptation for keeping water loss to a minimum; broad leaves lose water by evaporation, but evaporation is kept to a minimum in plants with only spines.

During the last ice age, the hot, dry climate gave way to more temperate conditions. Temperatures were cooler and precipitation was more abundant. This change brought a completely different look to the landscape, as the conifer woodlands expanded down the mountains and out onto the plains. The vegetation was slow to respond to the climatic changes at the end of the ice age, but the insect fossils show us that this change was both abrupt and early, perhaps as early as 12,500 yr B.P. The procession of climatic changes and biological responses to change kept right on going throughout the Holocene, and the "modern" biological communities did not become fully established until the last thousand years or even more recently.

As with some of the other regions we have examined in this book, the lasting impression to be taken away from Big Bend is one of the tenacity of life through a long history of changing environments. To be sure, the desert is a hard taskmaster for life in this region. But the land is not devoid of life. On the contrary, it is richly endowed with species that not only survive here, but thrive.

When the U.S. Congress set aside Big Bend National Park in 1944, one of the mandates given to the National Park Service was to restore the park to the condition it had enjoyed before the arrival of Europeans. This meant that all ranching, farming, and mining had to cease. The biota of the park had been severely affected by these activities, but since that time it has rebounded surprisingly well. Big Bend is a park that is not "on the way" to anywhere else; you do not pass by the entrance on your way from one urban center to another. But if you want to see the Chihuahuan Desert in action, this is the place to come.

Suggested Reading

Dick-Peddie, W. A. 1993. *New Mexico Vegetation, Past, Present, and Future.* Albuquerque: University of New Mexico Press. 244 pp.

Elias, S. A., and Van Devender, T. R. 1990. Fossil insect evidence for late Quaternary climatic change in the Big Bend region, Chihuahuan Desert, Texas. *Quaternary Research* 34:249–261.

Van Devender, T. R. 1977. Holocene woodlands in southwestern deserts. *Science* 198:189–192.

Van Devender, T. R. 1985. Climatic cadences and the composition of Chihuahuan Desert communities: The Late Pleistocene packrat record. In Diamond, J., and Case, T. J. (eds.), *Community Ecology.* New York: Harper & Row, pp. 285–299.

Van Devender, T. R. 1990. Late Quaternary vegetation and climate of the Chihuahuan Desert, United States and Mexico. In Betancourt, J. L., Van Devender, T. R., and Martin, P. S. (eds.), *Packrat Middens: The Last 40,000 Years of Biotic Change.* Tucson: University of Arizona Press, pp. 104–133.

Van Devender, T. R., and Bradley, G. L. 1990. Late Quaternary mammals from the Chihuahuan Desert: Paleoecology and latitudinal gradients. In Betancourt, J. L., Van Devender, T. R., and Martin, P. S. (eds.), *Packrat Middens: The Last 40,000 Years of Biotic Change.* Tucson: University of Arizona Press, pp. 350–362.

Wauer, R. H. 1980. *Naturalist's Big Bend.* College Station: Texas A&M Press. 149 pp.

Wells, P. V. 1966. Late Pleistocene vegetation and degree of pluvial climatic change in the Chihuahuan Desert. *Science* 153:970–975.

9

CONCLUSION

What Can We Learn from the Past?

My goal in writing this book has been to convey some of the sense of wonder and fascination for the southwestern wilderness as preserved in its national parks. But more than that, I hope that you have come to understand that what can be seen there today, however awe-inspiring, is only the latest page in the very long book of life written for this great region. That book is strange and unfamiliar to most people, since it is written in packrat middens, sediments, and cave deposits. Because it is unfamiliar, most have overlooked it. But once you have been introduced to it, the ancient past has a lot to offer.

For this historical perspective (some might call it the prehistoric perspective), you begin to see nature in a new way, by gaining an appreciation for the fact that modern ecosystems and landscapes have been shaped by past events. As I mentioned in Chapter 1, the current crop of plants and animals in biological communities are just the most recent biological actors to appear on stage for one brief act of a very long play. Furthermore, the combination of species that occurs in a specific ecosystem may or may not be the one best suited for that particular environment. Some will probably be gone in a few centuries. Newcomers, better adapted to the system, will squeeze them out. The scene appears stable to our eyes, but in reality it is constantly shifting. So when you come to the parks of the Southwest, stand on a high promontory, look over the landscape, and ask yourself, "How did it come to be this way?"

The Fragility of the Parks' Modern Ecosystems

Mesa Verde, Chaco Canyon, Canyonlands, Grand Canyon, and Big Bend national parks preserve patches of wilderness ranging from small to relatively large. The National Park Service has been given a dual mandate that often seems contradictory: to preserve these wilderness areas and at the same time to provide public access to them. How can you preserve something as vulnerable as a desert ecosystem while you facilitate its invasion by millions upon millions of visitors? One of the chief ways of overcoming this hurdle is to teach people to tread softly in the wilderness, to treat it with respect, and to give it the consideration it deserves. After all, it is the wilderness—that sense of wild, unspoiled nature—that endears these parks to their visitors.

We are always in danger of loving these parks to death. This happens when we treat the wilderness as just another theme park or tourist attraction, rather than as the precious, rare commodity it really is. Having visited Europe, where there is essentially no wilderness left at all (that which appears to be wilderness has actually been shaped and managed by people for many centuries), I appreciate our southwestern wilderness regions all the more. I hope that the descriptions in this book of how things came to be the way they are will add to your sense of wonder and your appreciation of these magnificent landscapes. But the bottom line is this: we must be careful with them, or they won't last very long.

Where Do We Go from Here?

I have described in this book how the ecosystems of the American Southwest follow cycles in response to climate change brought on by glacial and interglacial periods. This is true up to a point, but it must be emphasized that the current interglacial is different from the many that preceded it, because of the presence of mankind.

The growing human population is using up the world's resources at an alarming rate. We've reached the steep part of the growth curve, the part where population doubles every few decades. In the western states, we talk about "sustainable growth," but this term really contradicts itself. No growth of population can be sustained indefinitely. Sustainable *negative* growth is needed. We have now managed to alter the natural world to the point at which natural biological cycles have been seriously interrupted. In the ancient past, these cycles tracked the course of climate change. The Earth's biota may not be able to make the necessary responses as the climate continues to change, now that people have all but eliminated the natural order in so many regions.

Paleoecology has taught science many lessons that could not have been learned in any other way. Packrat midden fossils from the Southwest have shown us that this region looked completely different in the Pleistocene than it does now. The vegetation zones shifted downslope, some species were dominant then that are only marginal today, and vice versa. Concepts of southwestern ecosystem management must take into account the fossil evidence showing that piñon-juniper woodland growing in semiarid landscapes does not bounce back rapidly after the trees have been felled. Biogeographers must look into the fossil record to understand the comings and goings of species; this may mean that terms like *endemic species* will be used far less often, as the history of species' distributions shifts from speculation based on modern data to factual data from the fossil record. The field of population genetics has moved from the laboratory to the rockshelter, as fossil data from packrat middens becomes incorporated into theories of species' longevity and population dynamics. This is one of the most exciting aspects of paleoecological research: it spans the gap between the past and the present and draws the two realms (and the scientists who study them) together into a meaningful continuum.

We are indeed fortunate that, in the southwestern region, many elements of the natural ecosystems have remained intact, at least in fairly large patches here and there. Let us do all that we can to preserve as much of the Southwest in as natural a state as possible. This conservation ethic ranges from letting predators back onto the Kaibab Plateau to preserving desert regions so that their inhabitants will have a place to live. We need to take stock of the situation from a truly long-term perspective. It will show us that, like the Paleoindians, we are just passing through the land. We may enjoy this region for a season, but it is the product of thousands of years of environmental change, preceded by millions of years of mountain building, canyon cutting, and shifting of the ranges of plants and animals. We do not want to be known as the people who brought an end to the wilderness in the Southwest, but because of our technology and the size of our populations, we now have the capability of ruining what remains of the wilderness. This does not have to happen, but we will have to work hard to prevent it from happening.

GLOSSARY

Acidic sediments Sediments that contain more hydrogen ions than hydroxyl ions; sediments with a pH less than 7.0. Examples include some organic-rich soils and peats.

Alkaline sediments Sediments that contain more hydroxyl ions than hydrogen ions; sediments with a pH greater than 7.0. Examples include marls or soils rich in calcium carbonate.

Arroyo In arid and semiarid regions of the Southwest, a small, deep, flat-floored channel or gully of an ephemeral or intermittent stream, usually with vertical or steeply cut banks.

Atlatl A device for throwing a spear or dart that consists of a rod or board with a projection (a hook or thong) at the rear end to hold the weapon in place until released.

Bering Land Bridge Unglaciated regions of eastern Siberia, Alaska, the Yukon Territory, and the continental shelf between Alaska and Siberia that remained above sea level for large parts of the Pleistocene.

Beta (β) particle A high-energy electron given off by radioactive decay.

Burin A stone tool having the blade ground obliquely to a sharp point.

Catchment basin A basin that accumulates the runoff of precipitation from a watershed.

Chitin A nitrogen-containing polysaccharide (carbohydrate) compound that forms the hard, outer layer in the skeletons of insects and other invertebrates.

Coprolite Fossilized excrement of animals.

Cordilleran Ice Sheet A Wisconsin Glaciation ice sheet that covered most of western Canada, extending south into Washington, Idaho, and Montana.

Coursed masonry A continuous level or layer (*course*) of stones or bricks laid throughout the length of a wall.

Cyclotron An electromagnetic machine that accelerates high-energy particles (e.g., protons and electrons) in a circular path; the particles approach the speed of light.

Desert scrub A type of desert vegetation dominated by shrubs and cacti. Depending on the geographic region, these include creosote bush, sagebrush, saltbush, tarbush, and yucca.

Diatoms A single-celled, microscopic alga characterized by cell walls reinforced with silica.

Ecosystem A biological community, including all its component plants, animals, and other organisms, together with the physical environment, forming an interacting system.

Exoskeleton The external skeleton of insects.

Fluvial sediments Sediments laid down by running water (streams and rivers of all sizes).

Glacial moraine A mound or ridge of unsorted glacial debris (silt, sand, gravel, cobbles, and boulders), deposited by glacial ice in a variety of landforms.

Herbivore An animal that eats plants.

Holocene epoch An epoch of the Quaternary Period, spanning the interval after the last glaciation (10,000 yr B.P. to recent).

Interglacial A long interval between glaciations in which the climate warms to at least the present level.

Interstadial A relatively warm climatic episode during a glaciation, marked by a temporary retreat of ice.

Ion An electrically charged atom or group of atoms. An atom with a high affinity for electrons may acquire an electron, thus becoming negatively charged.

Isotope A variety of an element. Isotopes of an element differ from one another in the number of neutrons contained in the atom's nucleus.

Jet stream A strong, narrow current of air in the upper troposphere or stratosphere, thousands of kilometers long and hundreds of kilometers wide.

Laurentide Ice Sheet An ice sheet covering most of eastern and central Canada and the northeastern and north-central United States during the Wisconsin Glaciation.

Macrofossil The macroscopic (easily seen by the naked eye) remains of an ancient organism.

Marl A sediment deposited in lakes, composed mainly of calcium carbonate, and mixed with clay or silt.

Megafauna Animal species whose adults are heavier than 40 kg (88 lb).

Mesic Of or pertaining to environmental conditions that are characterized by medium levels of moisture supply.

Microfossil The microscopic (poorly seen or invisible to the naked eye) remains of ancient organisms.

Microsculpture Microscopic sculpture, including striations, punctures, and meshes, on the surface of, for example, insect exoskeletons, mollusk shells, or seed coats.

Packrat midden An accumulation of dried packrat feces and urine, usually containing plant macrofossils and other fossil remains, preserved in dry caves and rock shelters.

Paleoecology The study of the relationships between ancient organisms and their environments.

Palynology The study of fossil and modern pollen.

Parkland A type of forest in which the trees are widely spaced over a landscape covered with herbaceous vegetation.

Periglacial environments Environments at the immediate margins of glaciers and ice sheets, greatly influenced by the cold temperature of the ice.

Permafrost Permanently frozen ground, found in arctic, subarctic, and alpine regions.

Petroglyph Rock art produced by scratching or scraping into the surface of a rock, leaving an image made by the indentation. In the American Southwest, this was often done by scraping off the layers of desert varnish, a dark coating on rock surfaces.

Photosynthesis The synthesis by plants of carbohydrates from carbon dioxide and water by means of chlorophyll, using light as a source of energy.

Pictograph A painting made on a rock surface.

Piedmont glacier A mountain glacier which flows out from the mountain front, onto the adjacent plains (the piedmont region).

Pleistocene epoch An epoch of the Quaternary Period, spanning the interval from 1.7 million years ago to 10,000 years ago. The Pleistocene is characterized by a series of major glaciations.

Pluvial lake A lake formed in a period of exceptionally heavy rainfall; specifically, a lake formed in the Pleistocene epoch during a time of glacial advance, and now either extinct or greatly reduced in size.

Projectile points Sharp, pointed heads of stone or other material, attached to a shaft to make a projectile that is thrown or shot as a weapon. These include spearheads, arrowheads, and darts.

Proxy data In Quaternary studies, data from, for example, fossil organisms, sediments, and ice cores used to reconstruct past environments; proxy data serve as a substitute for direct measurements of such phenomena as past temperatures, precipitation, and sea level.

Quaternary Period The second period of the Cenozoic era, following the Tertiary and spanning the interval from about 1.7 million years ago to the present.

Stratigraphic column The arrangement of layers of sediments (strata) in geographic position and chronologic sequence.

Taphonomy The process by which organisms become preserved in the fossil record; also the study of that process.

Travertine A dense, finely crystalline, massive limestone deposit, white to tan in color, formed by rapid precipitation of calcium carbonate from solution in surface water or groundwater, or by evaporation from the mouth of a hot spring.

Treeline The altitudinal limit of tree species in mountain regions and the latitudinal limit of tree species at high latitudes.

Unconsolidated sediments Sediments with particles not cemented together or turned to stone.

Water-lain sediments Sediments that are deposited in water.

Wisconsin Glaciation The last major glaciation in North America, spanning the interval from about 110,000 to 10,000 yr B.P.

Xeric Of or pertaining to a dry habitat.

INDEX